Sociology 4F

TABLE OF CON & ACKNOWLEDG.

PAGE

The student perspective on employability   1
   Tymon, A.
   Studies in Higher Education, 38.6
   © 2013 Routledge Chapman & Hall Inc.
   Reprinted with permission.

Undergraduates' perceptions of employer expectations   17
   DuPre, C. & K. Williams
   Journal of Career and Technical Education, 26.1
   © 2011 Omicron Tau Theta
   Open access

Combining the liberal and useful arts: Sociological skills in the global economy   29
   Finkelstein, M.S.
   The American Sociologist, 25.3
   © 1994 Springer-Verlag New York Inc.
   Reprinted with permission.

Idealists vs. careerists: Graduate school choices of sociology majors   47
   Spalter-Roth, R. & N. Van Nooren
   Idealists vs. careerists: Graduate school choices of sociology majors,
   Spalter-Roth, R. & N. Van Nooren
   © 2009 American Sociological Assn
   Reprinted with permission.

The future lives of sociology graduates   63
   Guppy, N. et. al.
   Canadian Review of Sociology, 54.2
   © 2017 John Wiley & Sons (Canada)
   Reprinted with permission.

Jobs, careers, & sociological skills: The early employment experiences of 2012 sociology majors   79
   Senter, M.S. et. al.
   Jobs, careers, & sociological skills: The early employment experiences of 2012 sociology majors, Senter, M.S. et. al.
   © 2015 American Sociological Assn
   Reprinted with permission.

Choosing my major and career: A sociological inquiry — 97
  Knoblock, J.
    Human Architecture: Journal of the Sociology of Self-Knowledge, 6.2
    © 2008 Ahead Publishing House
    Reprinted with permission.

Called to Account: The CV as an Autobiographical Practice — 107
  Miller, N. & D. Morgan
    Sociology, 27.1
    © 1993 Cambridge University Press - UK
    Reprinted with permission.

Students' perceptions of reference letters — 119
  Payne, B. et. al.
    College Student Journal, 40.4
    © 2006 Project Innovation
    Reprinted with permission.

Reflections on the academic job search in sociology — 127
  Cotten, S.R. et. al.
    The American Sociologist, 32.3
    © 2001 Springer-Verlag New York Inc.
    Reprinted with permission.

Social Capital Activation and Job Searching: Embedding the Use of Weak Ties in the American Institutional Context — 145
  Sharone, O.
    Work and Occupations, 41.4
    © 2014 Sage - Journals
    Reprinted with permission.

Strong ties, weak ties, or no ties: What helped sociology majors find career-level jobs? — 177
  Spalter-Roth, R. et. al.
    Strong ties, weak ties, or no ties: What helped sociology majors find career-level jobs?, Spalter-Roth, R. et. al.
    © 2013 American Sociological Assn
    Reprinted with permission.

Facebook fired: Legal perspectives and young adults? opinions on the use of social media in hiring and firing decisions — 193
  Drouin, M. et. al.
    Computers in Human Behavior, 46
    © 2015 Elsevier Science
    Reprinted with permission.

The Demand Side of Hiring: Employers in the Labor Market
Bills, D.B.
Annual Review of Sociology, 43
© 2017 Annual Reviews
Reprinted with permission.

The student perspective on employability

Alex Tymon*

Centre for Organisational Research and Development, Portsmouth Business School, University of Portsmouth, Richmond Building, Portland Street, Portsmouth, PO1 3DE, UK

> Despite ongoing debate about whether they can and should, most higher education institutions include the development of employability skills within their curricula. However, employers continue to report that graduates are not ready for the world of work, and lack some of the most basic skills needed for successful employment. Research into why this might be abounds from the perspectives of multiple stakeholders, including government, employers, higher education institutions and graduates. Interestingly though, the views of undergraduates, the recipients of this employability development, are not well known. This could be important, because learning theory tells us that motivation and commitment of learners is an essential prerequisite for effective outcomes. So the question is raised as to whether undergraduate students are engaged with employability skills development. This article reports on a study exploring the views of over 400 business studies, marketing and human resource management undergraduate students about employability. Findings suggest there is only limited alignment between the views of students and other stakeholder groups. There are differences between first, second and final year students, which could explain an observed lack of engagement with employability-related development. Some suggestions for improving engagement are made, alongside ideas on what can, realistically, be done within higher education institutions.
>
> **Keywords:** employability; graduate skills; development; proactive personality; engaged learning

Introduction

Despite ongoing differences in views amongst stakeholders on what employability is, whether it can be developed and, perhaps most heatedly, the role of higher education institutions in its provision, there is increasing pressure for all academic courses to include employability development. Evidence suggests that, although the provision of employability skills is not consistent, many universities are expending a great deal of effort on developing the employability of their students (Harvey 2005; Higher Education Funding Council for England 2003; Yorke 2004). Yet research continues to report that graduates do not have the skills needed for the modern workplace (Bowers-Brown and Harvey 2004; Cumming 2010; Heaton, McCracken, and Harrison 2008). In the UK, the 2008 survey by the Confederation of British Industry found that 48% of employers were experiencing problems filling jobs with appropriately skilled graduates. Branine (2008) reports on a survey of 700 UK-based employers, where more than 60%

---

*Email: alex.tymon@port.ac.uk

© 2013 Society for Research into Higher Education

mentioned problems of poor-quality graduates in terms of their employability skills. These statistics could imply that this is a UK issue alone, and, as Jackson (2009) points out, there is significantly more research and survey data on graduate employability deficiencies in the UK than elsewhere. However, the demands of economic globalisation on higher education institutions across the world are recognised by many scholars (Cumming 2010; Jackson 2009; Kreber 2006). Kreber identifies employability as a key graduate outcome across multiple countries, and Jackson suggests that industry and governments worldwide would welcome effective ways to bridge graduate skills gaps. So the amount of UK data could be due to other factors; for example, the recent changes to university funding in the UK may have given the issue a higher profile for UK stakeholders. Either way, Cumming states: 'A dominant theme emerging ... is that many graduates lack appropriate skills, attitudes and dispositions, which in turn prevents them from participating effectively in the workplace' (2010, 3).

The nature of these skills can be derived from a study by Archer and Davison (2008). They found that communications was consistently ranked as the primary skill sought by employers, but in terms of employers' satisfaction with the quality of communication skills demonstrated by graduates, it ranked only sixteenth. Team working and integrity were ranked second and third in terms of importance, but only seventh and ninth in terms of satisfaction for employers. The authors go on to say: 'It appears that while many graduates hold satisfactory qualifications, they are lacking in the key "soft skills" and qualities that employers increasingly need in a more customer focussed world' (2008, 8).

This article aims to explore some of the myriad reasons why this situation may exist, including: the difficulties of defining the term 'employability' along with the transferable skills which it may include; and the extent to which employability matters to the various stakeholder groups. The article questions whether these skills can actually be developed, and if so, whether higher education institutions are the appropriate place to do so. The article then discusses what appears to be a less well-researched area: to what extent undergraduate students are engaged with the concept of employability, and are they willing and able to benefit from employability skills development in higher education institutions? This discussion is based on data collected from over 400 UK-based business students during October 2009.

What is employability?

It is suggested that one potential problem with trying to develop employability is a lack of coherence about what is meant by the term itself and the subsequent measurement of it. Most authors agree that employability is complex and multidimensional and warn against being simplistic when trying to define it (Harvey 2005; Holmes 2006; Rae 2007). Hugh-Jones, Sutherland, and Cross (2006) suggest that part of this complexity is because it can be viewed from three different perspectives: that of the employer, the student, and the higher education institution. Further complexity is noted by Rothwell and Arnold (2007), who highlight that employability can be viewed as having both internal and external dimensions. However, similarities exist across many of the definitions used, which resonate with that of Yorke, who defines employability as:

> a set of achievements, skills, understandings and personal attributes, that make graduates more likely to gain employment and be successful in their chosen occupations, which benefits themselves, the workforce, the community and the economy. (2004, 410)

This definition and others (e.g. Harvey 2005; Little 2001; Pool and Sewell 2007) distinguish between the *ability* to get a graduate-level job and *employment*, potentially due to the external factors reported by Rothwell and Arnold (2007). Thus, as Wilton states: 'it is possible to be employable, yet unemployed or underemployed' (2011, 87). This difference, between employment rates and employability, makes measurement of the concept challenging. Currently, most stakeholder groups use statistics from graduate destinations surveys to measure employability, whereas what these provide is a limited snapshot of employment. Yorke's definition also places focus on quality and sustainability of employment, a theme mirrored by others (e.g. Fugate, Kinicki, and Ashforth 2004), who stress the future-oriented nature of employability, with a need for adaptability and transitioning in future career market places.

Most definitions recognise that employability requires the possession of skills, but also personal attributes, which are aligned to personality theory. This link to personality theory, along with the qualitative nature and future orientation of the definitions, presents yet further challenges to measurement of the concept of employability.

What are the skills and personal attributes that make up employability?

Many terms are used in the literature to describe transferable skills and attributes: '"generic skills", "attributes", "characteristics", "values", "competencies", "qualities" and "professional skills"' (De La Harpe, Radloff, and Wyber 2000, 233). Along with each term there is often a proposed framework or list, some stretching to as many as 80 items. Table 1 provides a comparison of six such frameworks from numerous different perspectives: Kreber (2006) summarises a list of what universities should provide, derived from the World Conference on Higher Education; thus she suggests it has considerable agreement across counties. Andrews and Higson's (2008) list was synthesised from multiple sources as a basis for interviews in four European countries with both employers and graduates. Abraham and Karns (2009) show competencies from both an employer and business school perspective in the United States; the top 10 in each category are listed. Archer and Davison (2008) provide a UK employer perspective, whilst Cumming (2010) cites an Australian government perspective.

Table 1 indicates some agreement on the skills and attributes linked to employability, both amongst the different stakeholders and internationally, with communication/ interpersonal skills and teamwork appearing in all lists (see items in bold). However, there is less agreement on other items, and perhaps this is why authors such as Harvey (2005) and Yorke (2006) urge caution when assuming that there is agreement on what employability is. There are many examples where frameworks differ. Notably, these skill and attribute divergences are not confined to those between separate groups of stakeholders, as there is evidence to show that views also differ within groups of stakeholders. Differences between the views of *graduates* from the UK, Europe and Japan were indicated by Little and contributors (2003). Differences between *academics* across different higher education institutions and even within the same institution have been noted by Barrie (2007). The lack of shared understanding of skills, or attributes, has perhaps been best explored in relation to *employers* as a group of stakeholders. Little (2001) raised the issue of whether employers behave rationally when recruiting graduates and suggested evidence to the contrary, a view supported by Brown, Hesketh, and Williams (2003). According to Moreau and Leathwood, 'Employers may want, for example, someone who is strong and decisive, but they will inevitably read these qualities differently in different applicants' (2006, 319). This suggests that the three

Table 1. Comparison of employability frameworks.

Kreber (2006, 5) Multiple countries – competencies higher education institutions should provide.	Andrews and Higson (2008, 413) Employer and graduate perspectives: multiple sources.	Abraham and Karns (2009, 352)		Archer and Davison (2008, 7) Employers in the UK.	Cumming (2010, 7) Government in Australia.
		Top 10 competencies identified by businesses in the USA	Top 10 competencies emphasised in the business school curriculum in the USA		
• *Be able and willing to contribute to innovation and be creative* • Be able to cope with uncertainties • *Be interested in and prepared for lifelong learning* • **Have acquired social sensitivity and communicative skills** • **Be able to work in teams** • Be willing to take on responsibilities • Become entrepreneurial • Prepare themselves for the internationalisation of the labour market through an understanding of various cultures • Be versatile in generic skills that cut across disciplines • Be literate in areas of knowledge forming the basis for various professional skills, for example, in new technologies	• *Professionalism* • *Reliability* • The ability to cope with uncertainty • Ability to work under pressure • Ability to think and plan strategically • **Capability to communicate and interact with others, either in teams or through networking** • **Good written and verbal communication skills** • Information and communication technology skills • *Creativity and self-confidence* • Good self-management and time-management skills • *A willingness to learn and accept responsibility*	• **Communication skills** • Problem solver • *Results oriented* • Interpersonal skills • Leadership skills • Customer focus • *Flexible/adaptable* • **Team worker** • *Dependable* • Quality focussed	• **Communication skills** • Problem solver • **Team worker** • Leadership skills • Technical expertise • Interpersonal skills • Business expertise • *Hard worker* • *Results oriented* • *Dependable*	• **Communication skills** • **Team-working skills** • *Integrity* • Intellectual ability • *Confidence* • *Character/personality* • Planning and organisational skills • Literacy (good written skills) • Numeracy (good with numbers) • Analysis and decision-making skills	• **Communication** • **Teamwork** • Problem solving • *Initiative and enterprise* • Planning and organising • Self-management • *Learning* • Technology

Highlighted in bold = commonly cited items which appear in all frameworks.
Highlighted in italics = attributes with clear links to personality traits.
Highlighted by underlining = attributes potentially linked to proactive personality.

different perspectives mentioned by Hugh-Jones, Sutherland, and Cross (2006) could be significantly expanded.

In addition, any apparent agreement on skills, or attributes, is amongst a list of *labels* and not a detailed examination of what these mean to the individuals, or groups, concerned (Holmes 2006). For example, do 'communication skills' or 'team working' or 'flexibility' mean the same to any two stakeholders at the same time? According to Jackson, 'Empirical studies on graduate employability liberally adopt different terms for competencies, resulting in confused findings' (2010, 29), which is a concern if these studies are then used to inform policy or practice.

Can employability be developed and, if so, how?
Can skills be developed?
Skills are defined as: 'any component of the job that involves doing something' (Harrison 2003, 269), and include manual, diagnostic, interpersonal or decision-making skills. Along with knowledge, skills development is well documented in learning, training and development literature. Although it is recognised that some skills are more difficult to develop than others, there is agreement that skills can be trained or, at least, developed.

Can personal attributes be developed?
Personal attributes, on the other hand, cross into the differential psychology literature on personality traits and other individual differences such as intelligence or cognitive ability. Personality can be defined as: 'the overall profile or combination of traits that characterise the unique nature of a person' (French et al. 2008, 97). To what extent personality traits are inherited, or can be developed, is still a contentious subject (Rutter et al. 1997). But, even if personality can be developed, it is recognised that these highly individual traits are deep rooted, with many formed at an early age. They determine success, performance, and career choices, and any development of them is a long-term and slow process (Woods and West 2010). Table 1 shows that many of the items fall into the category of personality traits (see items in italics). Woods and West tell us that managers are looking for personality as often as skills, saying they want 'employees who are reliable, dependable, able to work under pressure, creative and enthusiastic. All of these reflect personality characteristics' (2010, 71).

In the United States, this area of research has been linked to 'proactive personality', a term defined by Seibert, Kraimer, and Crant as 'a stable disposition to take personal initiative' (2001, 847). Erdogan and Bauer add: 'Rather than accepting their roles passively, proactive persons challenge the status quo and initiate change' (2005, 859). A growing body of literature has shown some important links between proactive personality and career success from two angles. First, proactive personality has been shown to make adjustment to work a quicker and smoother process, resulting in people reaching effective performer status faster and more easily (Seibert, Kraimer, and Crant 2001). Second, there is a link to the process of job search, with people high on proactive personality more likely to succeed in this self-driven activity (Brown et al. 2006). Amongst the items listed in Table 1 are traits that could be linked to proactive personality (see items underlined), so perhaps stakeholders need to be more realistic about what can be developed in the higher education curriculum. Villar and Albertin (2010) summarise the work of many authors when they suggest that the role of higher education institutions should be to encourage students to develop their proactive personality traits. They propose this is done by getting students to take more responsibility for their education through active participation in educational experiences and intentional investment in their own social capital. At the very least, this area deserves further research.

Are higher education institutions the best place to develop employability?

The advent of mass higher education seen in the last three decades, and related growth in the number of vocationally oriented courses offered, appears to have changed expectations for many stakeholder groups (Bowers-Brown and Harvey 2004; Wilton 2011). Certainly, there is an expectation from government and employers that higher education institutions have a responsibility to prepare graduates for the world of work (De La Harpe, Radloff, and Wyber 2000; Heaton, McCracken, and Harrison 2008). In response, higher education institutions continue to build employability into their programmes (Bowers-Brown and Harvey 2004; Fallows and Steven 2000; Harvey 2005). Data also show that the majority of graduates recognise that higher education institutions are trying to support the employability agenda (Doctorjob.com 2004; Wilton 2008). But the expectation that higher education institutions can, and should, develop employability is not universally shared.

Many authors maintain that employability is better and more easily developed outside of the formal curriculum (Andrews and Higson 2008; Ng and Feldman 2009; Rae 2007; Yorke 2004), with particular emphasis placed on employment-based training and experience. There is little doubt that employers and employers' organisations are probably best placed to provide this work based training and experience, which in the past they did. However, organisations are becoming increasingly reluctant to invest in developing the transferable skills of graduates due to economic pressures and beliefs about the lack of commitment from 'generation Y' employees (Jackson 2010), and so higher education institutions are expected to fill the gap and produce work-ready employees. Yet Cranmer (2006) concluded that there was no evidence to show that employability skills development within universities had any effect on employability, compared to employment-based training and experience, which had positive effects. Graduates themselves are aware of the power of work experience in developing employability skills, with as many as 90% saying, 'work experience was the best way to gain the skills they needed for work' (Doctorjob.com 2004, 2). In addition, students on degree courses that include a work placement (sandwich courses) are up to 14% more successful in finding graduate employment compared to non-sandwich course students (Harvey 2005), due to the high value placed on work experience by employers. Although this could also be due to the opportunities these students have had to develop contacts. But this evidence suggests a need to be realistic about the effectiveness of employability skills development within higher education institutions and whether they are the best place to try and do so.

Are higher education institutions able to develop employability?

In addition to the debate on whether higher education institutions are the right place to effectively develop employability skills, there is also the question of whether they are able to do so. Kreber (2006) points out the multiple pressures on higher education institutions which could make it harder for them to give increased focus to the employability agenda: competing in the research arena; increasing numbers of students and their diversity, implying they are less prepared for university; along with declining resources. Rae (2007) tells us that universities are independent enterprises competing for student numbers in order to secure income, and this has not encouraged them to consider employers' needs when planning courses. He suggests that this has led to an increase in the number of 'trendy courses' offered at the expense of more traditional courses which employers value.

Should higher education institutions develop employability?

Far more contentious and fundamental than whether higher education institutions can develop employability skills is the philosophical question of whether they should.

Education in its broadest sense has been shown to positively correlate with both fluid and crystallised intelligence, core task performance and citizenship performance (Ng and Feldman 2009), all of which can contribute to employability. For some academics this broad education experience is not only sufficient, but is a core principle of higher education. They believe that higher education institutions are not the place to train graduates for jobs; that this is the responsibility of employers. Bowers-Brown and Harvey (2004) refer to the concept of the 'elitists', who believe there is an over-emphasis on vocational subjects, which is not the role of universities. Moreau and Leathwood (2006) talk about the increased focus on skills development threatening academic freedom. Kreber adds to this: 'some critics caution that universities could far too easily lose sight of such traditional values as curiosity-driven research, social criticism and preparation for civic life' (2006, 7). Some academics object to the philosophical changes being forced on higher education institutions (Jackson 2009), which appear to have coincided with a documented shift in the motivation to study, away from intellectual discovery towards a more instrumental approach (Massingham and Herrington 2006). Cornford (2005) argues that government-created expectations that employers' demands should be immediately responded to is the root cause of many higher education issues.

So, it is by no means clear whether employability skills can be developed and, if they can, the best way to do so. It is also debated to what extent higher education institutions can, and should, be part of employability skills development. But even if we can answer these questions another one remains: does employability matter?

To what extent does employability matter?

As with the definition of employability, the extent to which it is judged to matter varies by stakeholder group.

The government perspective

The UK government has a long-standing interest in higher education and employability, not least because it is the principal funder via taxation income. In more recent years, this interest has become more overt. Graduate employability has become a key objective for government and a performance indicator for higher education institutions. This focus on employability demonstrates what Cornford describes as 'an exceptionally instrumentalist approach' (2005, 41), and Wilton calls 'an economic ideology of higher education' (2008, 143) replacing the former view of what higher education institutions are for. This suggests that *employment* matters to government but, as discussed earlier, this is not necessarily employability.

The employer perspective

Branine (2008) found that graduate employers are more interested in personal attributes and soft skills than degree classification, subject or university attended. This view is supported by the Confederation of British Industry (2008), with 86% of board executives putting skills and attitudes at the top of their list of demands; degree result was rated as important by 32% and university attended was rated as important by just 10%. Nevertheless, this is contradicted by other evidence. Research by Wilton (2011) confirmed findings from previous studies, by showing that new university students fared less well in the labour market than those from older universities. This could indicate that employers' actions may not be matching their words.

The graduate perspective

For many graduates the economic drivers are strong. They recognise the value of employability skills and that a degree on its own may not be enough (Moreau and Leathwood 2006; Tomlinson 2008). The number of students graduating in the UK has increased dramatically in the last two decades, more than doubling since 1991, which has potentially led to an over-supply of graduates who find it hard to start their careers (Branine 2008; Rae 2007). This is evidenced by an increase in graduate unemployment, increased competition between graduates, and higher levels of uncertainty about what graduates can expect from higher education (Chartered Institute of Personnel and Development 2006; Moreau and Leathwood 2006). Not surprisingly, the increase in the number of graduates has also changed employers' expectations. A degree, once a bonus or differentiator, is now almost seen as a prerequisite for a job, even in sectors which in the past would not have needed a degree at entry level (Brown, Hesketh, and Williams 2003; Tomlinson 2008). Graduates are increasingly aware that they need additional skills and attributes for career success.

The higher education institution perspective

From the higher education institution perspective, the argument is simple: league tables can affect student numbers, which in turn affects funding. Despite arguments about the correlation between employability skills development and actual employment, higher education institutions need good employment figures. Therefore, they need to continue investing in, and promoting, employability development.

Wider society perspective

There are also those who suggest that employability skills are vital to society in general, as they enable people to contribute to the wider social environment (Brown, Hesketh, and Williams 2003; Wilton 2008).

The missing perspective

The missing perspective is the view of current students. Because these students are the intended recipients of employability skills development, their views are important. Most textbooks on learning theory highlight the need for learner motivation and engagement with the process to ensure effectiveness (e.g. Gold et al. 2010). Yet, we know little about the extent to which employability matters to current students, and what employability is from their perspective. Do they have similar views to other stakeholders on what transferable skills, or attributes, might be necessary? Do they think employability can, and should, be learned? Anecdotal evidence suggests that, for some students, most notably first and second years, there is a lack of engagement with the concept. These observations are supported by the literature (Rae 2007), which mentions the lack of appreciation by students of employability skills development.

Three other potential sources of current student views have been uncovered. Moreau and Leathwood (2006) carried out a longitudinal study with 310 mixed-discipline undergraduates. For these students, from a post-92 university, employability began to emerge as an issue as the study progressed and some of their findings are relevant to this article. Rothwell, Herbert, and Rothwell (2008) examined the beliefs of 344 undergraduate business students about their chances of success in seeking a particular type of work. Their findings included that the university attended had little impact on their self-perceived employability, as opposed to subject choice, which was rated as the top influencing factor. A striking finding was the perception that their level of engagement with studying was the least important factor linked to their employability. However, these researchers did not overtly explore the term 'employability' and its importance, nor the skills or attributes it may comprise. Tomlinson (2008) looked at 53 undergraduates and their perceptions of the role higher education credentials would play in shaping their future labour market outcomes. These students believed that degree qualification had lost differentiation value, and that there was a need to develop their wider employability. However, this sample group was limited to final-year students; we do not know if their views were the same earlier in their university career, when employability development could have occurred.

Methodology

Data was collected from first, second and final year undergraduate students in one post-92 UK university. Students were majoring in business studies/business administration, human resources and marketing. The final-year sample included sandwich students, newly returned from placement, and non-sandwich students. This conservatively includes 50% of the sample population for first-year students, 65% of the population for second-year students and 5% of the population for final-year students. It is recognised that the sample size for final-year students is low, and therefore care has been taken in reporting results from this group. There are other obvious limitations with this sample, which are discussed at the end of this article. The predominant method of data collection was via focus groups, which allowed the gathering of collective views and the collation of a joint construction of meaning (Bryman and Bell 2007). The non-sandwich final-year data was collected via questionnaire.

The questions posed were:

(1) What is your understanding of the term employability?
(2) What, if any, are the core/transferable skills that might make up employability?

(3) Either: (a) For first year students: To what extent do you expect the university to support the development of your employability, and how? (b) For all other groups: How much does university support the development of your employability, and how?

(4) To what extent do you think employability matters?

Findings and discussion

General findings

The number of responses per student increased by year, which indicates an increased confidence in self-expression. Focus-group observers reported that first-year students were more hesitant about contributing and their participation was far from equal. Second-year students appeared more confident in expressing themselves, with double the number of responses of first-year students, but again there was evidence that participation was unequal. Final-year students were extremely confident in expressing their views, with 14 times the number of items mentioned than by first-year students. This increasing confidence is of interest. It could be deemed to be evidence of enhanced communication skills and self-confidence, which regularly appear in employability skills frameworks, and which may suggest that these skills have been developed over the academic years.

Questions 1 and 2: what is employability and the skills/attributes it may encompass?

There is some alignment between the views the students expressed and the literature on the definition of employability and the skills and/or attributes it may include. All years and groups agreed that employability involved possession of skills linked to the needs of employers. In line with the literature, communication skills and team working were most commonly cited. Planning and organising and information technology skills were also commonly mentioned, and these appear in some of the frameworks reviewed for this article (see Table 1). All groups and years also agreed that personal attributes were an inherent part of employability, with the most commonly mentioned being: flexibility, adaptability, hardworking, commitment and dedication. Again this shows some alignment with the literature.

There was less alignment with the longer-term, wider definitions of employability (Fugate, Kinicki, and Ashforth 2004; Rothwell and Arnold 2007). This could suggest that these students are more concerned with the instrumental or economic view of employability discussed by Cornford (2005) and Wilton (2008). The final-year students did show some awareness of employability in its widest sense, suggesting it was about 'ensuring future employment'. This supports the findings of Tomlinson (2008), whose final-year students did consider longer-term advantages for graduates over non-graduates.

This pattern of alignment with the literature by academic year was also seen in relation to the value of qualifications or degree classification. Less than 40% of first and second year groups mentioned qualifications or grades as being connected to employability, whereas for employers a degree has almost become a prerequisite (Brown, Hesketh, and Williams 2003). Perhaps this finding may go some way towards explaining an observed lack of concern about grades for many first and second year students, 'First-year results don't matter' being a comment anecdotally heard. However, views of final-year students on the worth of qualifications were similar to the literature, with comments such as 'Education is number one', and 'A degree is standard, you need more'. This confirms the findings of Tomlinson (2008), who reported that final-year students placed a great deal of importance on their qualifications and believed employers would use degree classification as a way to differentiate between increasingly large pools of graduates.

The importance of experience also revealed differences between the years. Final-year students stated that experience was essential, agreeing with the studies by both Moreau and Leathwood (2006) and Tomlinson (2008). This indicates an understanding of employers' wants (Cranmer 2006; Doctorjob.com 2004; Ng and Feldman 2009; Yorke 2004). However, experience was only mentioned by half of the first and second year groups. This may indicate that many of these students do not have an informed understanding, or awareness, of what employers are looking for at this stage of their education.

Question 3: development of employability skills in the university

Echoing the findings from Moreau and Leathwood (2006), top of the list on university support, for all groups and years, was the placement opportunity. This was closely followed by the (faculty) placement office's curriculum vitae writing support and the (central) careers and jobs centre. This implies that getting a placement and gaining experience was well recognised as a university support. A final-year student commented, 'The placement was the main reason for picking this degree', and responses for final-year students to this question were congruent with their other answers. Interestingly, this was not always the case for first and second year students, whose answers presented contradictions. The placement and job search support were rated as most important in answer to question three, and yet experience was not rated highly as a key employability skill in response to question two. This raises the question: do these students really value placements and work experience (at this time), or is this just 'lip service'? Perhaps this may explain the anecdotal, observed and researched lack of engagement with placement-related activities (Rae 2007).

All groups and years mentioned embedded activities, such as presentations, group work and meeting deadlines, designed to develop skills/attributes such as communications, confidence, teamwork and self-management. However there was less emphasis placed on these embedded activities compared to placements and work experience. The lack of emphasis on embedded activities could be due to the nature of them: do students recognise that they are designed to develop employability skills? However, this is a research stream beyond the scope of this article.

Another area of interest is the low perceived value of student-driven activities, such as involvement in societies, volunteering and other extra-curricular opportunities. These were mentioned by less than half the first and second year groups and not mentioned at all by final-year sandwich students. Conventional wisdom would view these as examples of demonstrating employability skills, and Tomlinson (2008) found that his final-year students believed that extra-curricula activities, such as societies and sports, were important. However, the evidence to support the value of these is mixed. For example, in respect of volunteering, Konidari (2010) found that the predominant reason for students carrying out volunteer work was to enhance their career and curriculum vitae. However, whilst students self-report that volunteering has improved their skills and employability, there is little empirical evidence to show that it actually achieves this aim (Hill, Russell, and Brewis 2009). To quote Holdsworth and Quinn:

> While there are subjective data on how students feel they have benefited from volunteering and in many cases students do get jobs through volunteering (e.g. youth and conservation work), the absence of a control group means that statistically the case for employability is not proven. (2010, 123)

So, perhaps our students are right to ignore our suggestions that they develop their employability through volunteer work. But this does indicate a need for further research.

Question 4: to what extent does employability matter?

All students said employability mattered a 'great deal' or 'massively', but with focus on getting a job, any job, as opposed to employability in its wider sense, as discussed earlier. Comments included: 'There is no point in university without employability' and 'It can put you above the rest, competition is fierce'. The majority of first and second year groups went no further with this question, which may indicate that the topic is not really important to them at this stage.

For those who did expand upon why employability matters, reasons tended to be individually and instrumentally focused: 'job security', 'better pay', 'increased choice of jobs'. A small minority of groups went on to suggest that employability may improve quality of employment, with statements such as: 'It will give you a more enjoyable career' and 'It helps you plan your life and shows your development needs'. This suggests that, for only a small number of students, employability may be a wider and more valuable concept than employment. Very few groups mentioned the benefits to others, such as employers, higher education institutions, taxpayers and society in general. The lack of expansion on this question, for first and second year students at least, leads one to consider whether or not they really do believe that employability matters, and are therefore engaged with the development of employability skills.

Conclusions

This article set out to progress the discussion about the complex topic of graduate employability, most notably in the area of undergraduate engagement with the concept.

Whilst recognising that there is no universally accepted definition of employability, the views of most of these students are narrow in comparison to the literature. They seem to believe that employability is a short-term means to an end, being about finding a job, any job, or employment. Many of the literature definitions take a much wider stance, suggesting that employability should be more concerned with longer-term quality and sustainability of graduate-level employment. The more instrumental view of employability seems to correlate with the views of current and more recent governments, evidenced by the simplistic way in which employability is measured through employment statistics.

More alignment between student views and the literature was found in terms of the skills and personal attributes associated with employability. The most commonly cited skills were communication, team working, information technology, and planning and organising. Personal attributes agreed upon included flexibility, adaptability, hardworking, commitment and dedication. However, it should be remembered that there is no universal agreement on the content of employability frameworks, either between or within stakeholder groups. Further, any agreement is just between 'labels', with little evidence to suggest that any of the interested stakeholders, including undergraduate students, share a common understanding of these terms. Nor is there concurrence about how they are assessed.

It is also unclear whether many of these skills and attributes can be developed in practice and, if so, what the role of higher education institutions should be. Putting aside the arguments about whether higher education institutions are able, willing or designed to develop employability, there is evidence to suggest there are alterative options which may be more appropriate.

Skills can be developed and are embedded in the curriculum, but many first and second year students appear to lack engagement with these activities. This must reduce their motivation to learn and inevitably impact on successful development. Higher education institutions could make improvements in this area, perhaps by increasing awareness of employability in its wider sense and the benefit to students of their engagement with the concept and/or perhaps by making skills development activities more overt. As individual benefits were clearly the main reason why students thought employability mattered, this could be a feasible objective, even if it does pander to the instrumental view of employability.

Personal attributes are more complex, with many falling into the category of proactive personality. Planned and explicit development of these is possibly outside the capability and remit of higher education institutions. Student-driven activities may be a way to develop proactive personality, but only a minority of these students recognised student-driven activities as a useful activity to develop employability. As an interesting aside, it could be that students who do commit to self-driven activities may actually be already high on proactive personality. Perhaps the way forward here is to focus on raising awareness of what employers need or want in terms of personal attributes, promoting the message of Villar and Albertin (2010) of the need for students to become more actively involved and responsible for their education, investing in their own social capital. Providing students with a better understanding of how student-driven activities can develop and/or demonstrate proactive personality could be a practical step.

Promotion of work-based training and experience may need to be reconsidered. There is clear evidence that these are the best techniques for the development of many employability-related skills and personal attributes. However, first and second year students may require more help to see the benefits of these activities, as their conflicting answers raised questions about their real engagement with the concept. Additionally, although experience is highly attractive to employers, there seems to be an increasing reluctance for them to supply development in transferable skills. This is certainly a theme which deserves further exploration.

Finally, there is the possible lack of importance associated with qualifications or degree classification by first and second year students, which is at odds with other stakeholder groups, including final-year students. If we are to raise the engagement levels of students in their first two years, they need to recognise that employers do put emphasis on qualifications, and because of the laws of supply and demand, employers can afford to be selective about grades.

Limitations of this study and further research

Various authors have suggested that business students should be more interested in, and have a greater awareness of, employability as they have opted to study a vocationally oriented subject (Berman and Ritchie 2006; Jackson 2009; Parrott 2010). Therefore, these students could have a more informed perspective which may limit the potential for generalisation of the results.

Another limitation is the use of just one department in one post-92 university, meaning the results may not be representative. Rothwell, Herbert, and Rothwell (2008) showed there was little difference in student perceptions among three different pre-92 and post-92 universities. This, along with the reasonable sample size for first and second year students, should enable this data to make a useful contribution. However, further studies are recommended to validate the results, especially with final-year students.

According to Bryman and Bell (2007), there are limitations to qualitative data collection methods such as focus groups. These include: control, group dynamics and data analysis issues, all of which may limit the value of the findings and generalisations made. However, it is hoped that this article will provide some useful insights for those committed to the employability agenda and will provide a basis for further work in this area.

Areas of further research abound and include: more detailed analysis of the skills and attributes frameworks to explore shared meaning; empirical evidence for the value of volunteering and other student-driven activities; the discrepancy between students saying placements were of number one importance, but not rating experience highly as an employability element. One further research area that springs to mind is to what extent could proactive personality be a 'chicken and egg' situation? Simply put, are students who are high on proactive personality more likely to be involved in student-driven activities, finding placements and skills development activities embedded in the curriculum? It would be interesting to assess for levels of proactive personality at an early stage, and then relate this to their answers to the research questions used in this study, and their subsequent performance at and involvement in university.

References

Abraham, S.E., and L. Karns. 2009. Do business schools value the competencies that businesses value? *Journal of Education for Business* 84, no. 6: 350–56.

Andrews, J., and H. Higson. 2008. Graduate employability, 'soft skills' versus 'hard' business knowledge: A European study. *Higher Education in Europe* 33, no. 4: 413–22.

Archer, W., and J. Davison. 2008. *Graduate employability: What do employers think and want?* The Council for Industry and Higher Education. http://www.cihe-uk.com (accessed June 18, 2011).

Barrie, S.C. 2007. A conceptual framework for the teaching and learning of generic graduate attributes. *Studies in Higher Education* 32, no. 4: 439–58.

Berman, J., and L. Ritchie. 2006. Competencies of undergraduate business students. *Journal of Education for Business* 81, no. 4: 205–9.

Bowers-Brown, T., and with L. Harvey. 2004. Are there too many graduates in the UK? A literature review and an analysis of graduate employability. *Industry and Higher Education* August: 243–54.

Branine, M. 2008. Graduate recruitment and selection in the UK. A study of recent changes in methods and expectations. *Career Development International* 13, no. 6: 497–513.

Brown, D., R. Cober, K. Kane, P. Levy, and J. Shalhoop. 2006. Proactive personality and the successful job search: A field investigation with college graduates. *Journal of Applied Psychology* 91, no. 3: 717–26.

Brown, P., A. Hesketh, and S. Williams. 2003. Employability in a knowledge-driven economy (1). *Journal of Education and Work* 16, no. 2: 106–26.

Bryman, A., and E. Bell. 2007. *Business research methods*. Oxford: Oxford University Press.
Chartered Institute of Personnel and Development (CIPD). 2006. Survey report. Graduates in the workplace: Does a degree add value? http://www.cipd.co.uk/hr-resources/survey-reports/graduates-workplace-degree-add-value.aspx (accessed June 22, 2011).
Confederation of British Industry (CBI). 2008. Education and skills survey, 2008. http://educationandskills.cbi.org.uk/reports (accessed May 13, 2011).
Cornford, I.R. 2005. Challenging current policies and policy makers' thinking on generic skills. *Journal of Vocational Education and Training* 57, no. 1: 25–45.
Cranmer, S. 2006. Enhancing graduate employability: Best intentions and mixed outcomes. *Studies in Higher Education* 31, no. 2: 169–84.
Cumming, J. 2010. Contextualised performance: Reframing the skills debate in research education. *Studies in Higher Education* 1–15, iFirst Article.
De La Harpe, B., A. Radloff, and J. Wyber. 2000. Quality and generic (professional) skills. *Quality in Higher Education* 6, no. 3: 231–43.
Doctorjob.com. 2004. Are graduate employability initiatives worth it? Student survey. July 2004. http://aces.shu.ac.uk/employability/resources/doctorjobsstudentsurvey0604.pdf (accessed June 22, 2011).
Erdogan, B., and T. Bauer. 2005. Enhancing career benefits of employee proactive personality: The role of fit with jobs and organisations. *Personnel Psychology* 58: 859–91.
Fallows, S., and C. Steven. 2000. Building employability skills into the higher education curriculum: A university wide initiative. *Education+Training* 42: 75–82.
French, R., C. Rayner, G. Rees, and S. Rumbles. 2008. *Organisational behaviour*. Chichester: Wiley.
Fugate, M., A. Kinicki, and B. Ashforth. 2004. Employability: A psycho-social construct, its dimensions and applications. *Journal of Vocational Behaviour* 65: 14–38.
Gold, J., R. Holden, P. Iles, J. Stewart, and J. Beardwell. 2010. *Human resource development: Theory and practice*. Basingstoke: Palgrave Macmillan.
Harrison, R. 2003. *Learning and development*. Trowbridge: Chartered Institute of Personnel and Development.
Harvey, L. 2005. Embedding and integrating employability. *New Directions for Institutional Research* 128: 13–28.
Heaton, N., M. Mccracken, and J. Harrison. 2008. Graduate recruitment and development. Sector influence on a local market/regional economy. *Education+Training* 50, no. 4: 276–88.
Higher Education Funding Council for England. 2003. How much does higher education enhance the employability of graduates? Summary report 2003. http://www.hefce.ac.uk/pubs/rdreports/2003/ (accessed June 22, 2011).
Hill, M., J. Russell, and G. Brewis 2009. Young people, volunteering and youth projects: A rapid review of recent evidence. http://vinspired.com/uploads/admin-assets/datas/282/original/v-formative-evaluation-rapid-evidence-reveiw-Dec-2009-x-2pdf (accessed May 18, 2011).
Holdsworth, C., and J. Quinn. 2010. Student volunteering in English higher education. *Studies in Higher Education* 35, no. 1: 113–27.
Holmes, L. 2006. Reconsidering graduate employability: Beyond possessive-instrumentalism. Paper presented at the seventh International Conference on HRD Research and Practice across Europe, May 22–24, in in Tilburg, Netherlands.
Hugh-Jones, S., E. Sutherland, and A. Cross. 2006. The graduate: Are we giving employers what they want? Paper presented at the Teaching and Learning Conference, January 6, in Leeds.
Jackson, D. 2009. Profiling industry-relevant management graduate competencies: The need for a fresh approach. *International Journal of Management Education* 8, no. 1: 85–98.
Jackson, D. 2010. An international profile of industry-relevant competencies and skill gaps in modern graduates. *International Journal of Management Education* 8, no. 3: 29–58.
Konidari, S. 2010. Exploring impacts of volunteering on university students. Reach, student development and activities services. http://reach.londonmet.ac.uk/fms/MRSite/psd/StudServ/SDAS/Our-Research-Impact-of-volunteering.pdf (accessed May 18, 2011).
Kreber, C. 2006. Setting the context: The climate of university teaching and learning. *New Directions for Higher Education* 133: 5–11.
Little, B. 2001. Reading between the lines of graduate employment. *Quality in Higher Education* 7, no. 2: 121–30.

Little B. and contributors. 2003 International perspectives on employability. A briefing paper. http://www.palatine.ac.uk/files/emp/1260.pdf (accessed June 14, 2010).

Massingham, P., and T. Herrington. 2006. Does attendance matter? An examination of student attitudes, participation, performance and attendance. *Journal of University Teaching & Learning Practice* 3, no. 2: 82–104. http://ro.uow.edu.au/jutlp (accessed May 13, 2011).

Moreau, M.P., and C. Leathwood. 2006. Graduates' employment and the discourse of employability: A critical analysis. *Journal of Education and Work* 19, no. 4: 305–24.

Ng, T.W.H., and D. Feldman. 2009. How broadly does education contribute to job performance? *Personnel Psychology* 62: 89–134.

Parrott, G. 2010. Redesigning the first year business curriculum at the University of Bedfordshire. *International Journal of Management Education* 8, no. 2: 13–21.

Pool, L.D., and P. Sewell. 2007. The key to employability: Developing a practical model of graduate employability. *Education + Training* 49, no. 4: 277–89.

Rae, D. 2007. Connecting enterprise and graduate employability. Challenges to the higher education culture and curriculum? *Education+Training* 49, nos. 8/9: 605–19.

Rothwell, A., and J. Arnold. 2007. Self-perceived employability: Development and validation of a scale. *Personnel Review* 36, no. 1: 23–41.

Rothwell, A., I. Herbert, and F. Rothwell. 2008. Self-perceived employability: Construction and initial validation of a scale for university students. *Journal of Vocational Behaviour* 73: 1–12.

Rutter, M., J. Dunn, R. Plomin, E. Simonov, A. Pickles, B. Maughan, J. Ormel, J. Meyer, and L. Eaves. 1997. Integrating nature and nurture: Implications of person–environment correlations and interactions for developmental psychopathology. *Development and Psychopathology* 9: 335–64.

Seibert, S., M. Kraimer, and J. Crant. 2001. What do proactive people do? A longitudinal model linking proactive personality and career success. *Personnel Psychology* 54: 845–74.

Tomlinson, M. 2008. 'The degree is not enough': Students' perceptions of the role of higher education credentials for graduate work and employability. *British Journal of Sociology of Education* 29, no. 1: 49–61.

Villar, E., and P. Albertin. 2010. 'It is who knows you'. The positions of university students regarding intentional investments in social capital. *Studies in Higher Education* 35, no. 2: 137–54.

Wilton, N. 2008. Business graduates and management jobs: An employability match made in heaven? *Journal of Education and Work* 21, no. 2: 143–58.

Wilton, N. 2011. Do employability skills really matter in the UK graduate labour market? The case of business and management graduates. *Work Employment Society* 25, no. 1: 85–100.

Woods, S., and M. West. 2010. *The psychology of work and organizations*. Andover: Cengage.

Yorke, M. 2004. Employability in the undergraduate curriculum: Some student perspectives. *European Journal of Education* 39, no. 4: 409–27.

Yorke, M. 2006. Employability in higher education: What it is and what it is not. Learning and Employability Series 1. http://www.heacademy.ac.uk/resources/detail/ourwork/employability/employability336 (accessed June 22, 2011).

Undergraduates' Perceptions of Employer Expectations

Carrie DuPre, MA, Ph.D.
Kate Williams, MEd, MS
Clemson University

ABSTRACT

Research conducted by the National Association of Colleges and Employers (NACE) indicates that employers across industries seek similar skills in job applicants; yet employers often report finding these desired skills lacking in new hires. This study closes the gap in understanding between employer expectations and student perceptions regarding necessesary on-the-job skills. To accomplish this understanding, students' perceptions of the skills employers seek in new hires were measured and then compared against the NACE research of employers. Students' perceptions of their preparedness for these skills were also analyzed. Findings indicated potential implications for on-campus curricula and programs to help support studetns prepare for successful careers.

Introduction

Students equate the value of their postsecondary education with the level to which that education prepares them for their future careers, and traditionally aged college freshmen consider the quality of a career services office and the overall job placement rate critical when evaluating a college (Farrell, 2007). Indeed, 85 percent of students report entering college with a career in mind, and 37 percent of students admit that they would drop out of college if they thought their attendance would not help their job chances (Levine & Cureton, 1998). This connection between education and employment is exemplified in today's depressed job market, where there are more applicants for fewer jobs.

For 2010, the National Center for Education Statistics estimated that about 2.4 million graduates entered the competition for job (Petrecca, 2010), yet the Congressional Budget Office projected the unemployment rate hovered around 10 percent during the same time period (Sunshine, 2010). These 2.4 million graduates are also competing with the 15 million Americans are already looking for work, including unemployed graduates from previous years, laid-off workers, and struggling retirees (Simon, 2010; Rampell & Hernandez, 2010; Petrecca, 2010). Given this dire employment situation, the need to understand students' perceptions of employer expectations, while they are still in college and can still seek opportunities to acquire such skills, is significant. This need guided the inquiry behind this study, which asks two main questions: What are undergraduate students' perceptions of skills employers seek in new hires? And, based on their curricular and extracurricular activities, how prepared do students feel they are to demonstrate these skills after college?

Research Findings: Employer Desires and Student Perception

Researchers from the National Association of Colleges and Employers (NACE) found that high grade point averages are still necessary in job candidates, though high grades alone are not enough to secure a job (Koc & Koncz, 2009). Non-technical skills are required in today's jobs, though, given how articulate employers appear to be with colleges about what they want in college graduates, vocational researchers are surprised by existing gaps between what employers want and what graduates possess (Hannerman & Gardner, 2010; Koc & Koncz, 2009; Koc, 2010; *Occupational Outlook Handbook*, 2010). For example, employers are clear in wanting college graduates who can demonstrate "relevant work experience" (Koc, 2010; "*Occupational Outlook Handbook,*" 2010, p. 22); yet, even given students' expressed understanding of the importance of real-world skills, academics and researchers repeatedly illustrate the lack of real-world skills in freshly minted graduates. There also exists a well-documented gap between desired non-technical skills and the non-technical skills college graduates possess, including effective interpersonal skills (Garber, 2003). Across industries, employers also place high value on potential employees who graduate from collegiate institutions that teach beyond simple knowledge transmission into preparing graduates to be "successful and contributing members of today's global economy" (Hart, 2006, p.1).

According to employers, and as reported through NACE research, the most effective way for graduates to stand out among the crowd is to prove they possess outstanding credentials in a number of non-industry specific desired skill areas. The 2010 Job Outlook, an annual NACE publication, listed these desired characteristics as, in order of importance, communication skills, analytical skills, teamwork skills, technical skills (as related to major), and a strong work ethic (Koc & Koncz, 2009). These skill areas are referenced frequently in this study as the *top five desired skills*. NACE did not define each term for participants; therefore, in this study's data collection tool assessing student perceptions, these terms are also not defined. Whether students and employers have similar understandings of these terms is an assumption that could be examined in future research.

In the 2010 Job Outlook, the National Association of Colleges and Employers surveyed 219 United States employers to find that 49.7 percent list *communication skills* as the most sought skill in employees, while at the same time list *communication skills* as the most lacking skill found in new college graduates. Of the other highly sought skills, 28.9 percent found *initiative* lacking, 27.2 percent found *teamwork skills* lacking, 20.8 percent found *strong work ethic* lacking, 11 percent found *analytical skills* lacking, and 8.7 percent *found technical skills* lacking (Koc & Koncz, 2009). Researchers have also found that workers show a "persistent gap between the skills needed and the skills possessed" in the use of technology (Morgeson, Campion, & Levashina, 2009, p. 203), and that the effectiveness of written communication skills taught to business students is "questionable" (Pittenger, Miller, & Allison, 2006, p. 257).

Identification of Variables to Study

For this study, NACE research outlining the top five desired skills was used to craft survey questions asking students their perceptions of the level to which employers valued each skill. These were also the skills listed when asking students to rate the level to which students felt prepared to demonstrate these skills in the workplace. A review of literature, as outlined in the following paragraphs, identified two areas highly credited for providing students with necessary growth: extracurricular and academic programs. Based on the campus of study, numerous extracurricular and academic programs were listed for students to select as having participated in and to then evaluate if they felt such programs impacted their abilities to demonstrate the top five desired skills.

Literature on Extracurricular and Academic Programs

Little empirical research exists on the impact of participation in extracurricular activities on career choice (Pascarella & Terenzini, 2005), yet a considerable amount of career-related skills are developed in non-instructional settings (Heckman, 1999). In his seminal work, Chickering (1969) included career development under the "establishing purpose" vector in his analysis of how students develop in college. Student affairs scholars argue the value of extracurricular involvement in "improving the quality of the undergraduate learning experience" (Pomerantz, 2006, p. 177), noting that when students become actively involved in campus activities, they benefit from greater "student learning and personal development" (Pascarella & Terenzini, 1991, p. 36).

Regarding overall academic experiences, Pascarella and Terenzini's review of research conducted in the 1990s discovered a small yet consistent body of research "suggesting that the quality of effort or involvement students make in meeting the requirements of their formal academic program has an impact on their self-ratings of growth in career-related competencies and skills" (2005, p. 522). Tagg (2003) identified similar relevance of hands-on experience in obtaining a job by arguing for the value of students taking an active role in their education and continually connecting what they learn to the world around them. This learning paradigm puts high responsibility on the student, viewing education as something one engages in rather than something one obtains. Learning therefore becomes a process in which students rely on their personal perspectives when deciding what to believe while they "simultaneously share responsibility with others to construct knowledge" (Baxter Magolda and King, 2004, p. xviii), as would be accomplished in a team setting on the job.

Many researchers identify the internship as the activity that offers the greatest impact on students' career preparation, pointing to positive outcomes including growth in academic performance (Dundes & Marx, 2006/2007), obtaining job-related skills (Pascarella & Terenzini, 2005; Kim & Alvarez, 1995), making professional contacts (Bottner, 2010), and developing soft skills such as professional etiquette and communication (Walgran, 2010). Having an internship is also a factor in many companies' hiring practices; over the past years, employers report growing importance placed on hiring graduates with internship experience (Koc, 2010).

Conceptual Justification for Study

A vast body of literature assesses only what employers seek in potential employees or what employers see in their job applicants and new-hires. The focus of such research lies with the employer and what skills they desire and/or observe. Another body of research investigating students and future employment centers on the concept of "vocational self-concept," regarding the connection one's self concept has to one's vocational development (Super, 1963, p. 20). The importance of self concept on career development is an idea long recognized in the field of vocational psychology (Betz & Hackett, 2006), with a person's self concept even dictating career selection (Holland, 1997). Self-efficacy has also played a central role in understanding how a person's self concept impacts career preparation. More specifically, self-efficacy assumes that a student's level of self confidence to perform a task impacts the level of effort expended, perseverance in the face of adversity, and the belief that one could successfully execute a desired course of action (Bandura, 1997).

Literature like that highlighted above does not, however, address the gap between what employers seek in potential employees and what students perceive as important skills to obtain. In order to best prepare students for a competitive workplace, this gap must be addressed. The current study differs from past research by investigating student perceptions of the skills they think employers desire and the level to which they feel they are working toward obtaining these skills before college graduation. The focus is on perceptions of importance and participation in activities that help build skills. This data begins to address the gap between employer desires and student perceptions, as well as activities students perceive as fostering growth in skills that support their career preparation.

Method

Institutional Location and Participants

Undergraduate student perceptions were analyzed at apublic, four-year institution in the Southeastern United States. This location is particularly relevant to the study of career preparedness as the most optomistic views of hiring projections for the class of 2010 still predict that the Southeast will report a 10 percent decrease in hiring (Koc, 2010). Participants were recruited from general education courses with the intent of recruiting students who represented diverse academic backgrounds. The participants, all of whom were at least 18 years of age, represented varying demographics, years in school from freshman to senior, and pursuit of various degrees offered by the research instituion. More specifically, of the 125 participants, 58 percent were female and 42 percent were male. Regarding ethnicity, 82 percent of the sample were White, very near the 81.72 percent of the institution's undergraduate population that reports being White. The participants were at various stages of their undergraduate degree: 26 percent were first-year students, 33 percent were sophomores, 26 percent were juniors, and 14 percent were seniors. Although more participants studied business and the behavioral sciences (39 percent), all of the academic colleges were substantially represented, with 13 percent studying agriculture, 13 percent engineering and science, 15 percent liberal arts, and 20 percent health and education.

Research Instrument

A voluntary paper-and-pencil survey was distributed during class to obtain participant perceptions of how employers value the top five desired skills most sought by employers: analytical skills, communications skills, major (technical) skills, teamwork skills, and work ethic. They were then asked about thier perceptions of how participation in institution-related experiences prepared them for their careers. Students were asked which of six common extracurricular activies they completed and their perceptions of how those activities prepared them in the top five desired skills most sought by employers. Next, students indicated their participation in 14 common academic programs and their perceptions of how those activities prepared them regarding the same five skills. The activities listed in the survey were derived from the American College Professionals Association initiatives (1996), as well as a comparison to which activities were offered at the institution of study. Demographic information was also collected, including gender, ethnicity, class year, college in which degree is sought, and plans for the first year after graduation.

Results

Ranked Importance of and Perceived Preparedness of Top Five Skills

First, undergraduate students' perceptions of skills employers seek in new hires were analyzed. Students perceived work ethic (m=3.81, 37%) as most important. This was followed, in ranked order, by: communication (m=3.69, 29%); teamwork (m=2.84, 31%); analytical skills (m=2.58, 37%); and technical skills (m= 2.09, 56%). Second, participants were asked to rate how prepared they felt to use the top five desired skills in the workplace. Students were most confident of their work ethic (M=4.46, SD=.82) and teamwork skills (M=4.34, SD=.79) followed by communication skills (M=4.22, SD=.71). They were less sure of their analytical skills (M=4.02, SD=.77) and major skills (M=3.85, SD=.93). These perceptions did not significantly differ by gender or ethnicity, although they did significantly differ by class year, with juniors and seniors reporting greater perceptions of preparations than freshmen (see Table 1). In addition, students studying health and education felt significantly more prepared to use major skills than students studying business and behavioral sciences (mean difference=.48, p<.05). There were no differences between the other academic colleges in students' perceptions of preparedness.

Perceived Preparedness to Demonstrate Skills

Undergraduate students were presented with a list of 19 skills employers seek in new hires to analyze their self concept of preparedness related to using those skills on the job. This longer list, which encompasses the top five desired skills, includes many of the skills noted in vocational research, largely by NACE, to gain a more total picture of students' vocational self concepts (Koc, 2010; *Occupational Outlook Handbook*, 2010). Two of the students' perceived top three preparedness skills are also highly desired by employers: work ethic and teamwork. (The third skill with which students perceived a high level of preparedness was friendly/outgoing.) However, another two of the employers' top five desired skills fall in the bottom third of the students' perceptions, analytical skills and technical skills, illustrating a gap in the expectations between the two parties. (See Table 2 for the full results.)

The students' ratings of the remaining 13 skills were significantly different than the employers' ratings of importance, again illustrating a potential gap in expectations. It is important to note the direction in which the students' perceptions of their skills differ from the skills desired by employers; students' perceptions of their skills were significantly lower than the employers' ratings of importance for communication, analytical, technical, and computer skills, among others.

Impact of Extracurricular and Academic Involvement

This study also sought to identify relationships between extracurricular and academic activities and students' perception of how those activities contributed to their growth in the desired skill. To obtain meaningful correlations, both variables needed to show adequate variance in scores. Therefore, the activities that were either universally endorsed (participation in speech classes, writing classes, and major and non-major classes), or that had limited participation (varsity athletics, study abroad, and electronic portfolio completion), were eliminated from these analyses.

Of particular interest regarding the relationship between students' perceptions of their skills and the activities they completed was the significant positive correlation between analytical thinking skills and student participation in undergraduate research ($r=.27$, $p<.01$), study abroad ($r=.21$, $p<.05$), and internships ($r=.19$, $p<.05$). Significant positive correlations were found between communication skills and tutoring other students ($r=.21$, $p<.05$) and non-major courses ($r=.18$, $p<.05$). Students' skills in their academic major had significant positive correlations with taking major courses ($r=.29$, $p<.01$), undergraduate research ($r=.24$, $p<.01$), completing their university-required electronic academic portfolio ($r=.22$, $p<.05$), and participating in an internship ($r=.19$, $p<.05$). Interestingly, a significant positive correlation with skills in the major also emerged for non-major courses ($r=.19$, $p<.05$) and participation in student organizations ($r=.18$, $p<.05$). A significant positive correlation emerged between work ethic and participation in student organizations ($r=.21$, $p<.05$). Finally, a significant negative correlation was found for teamwork skills and Greek life participation ($r=-.24$, $p<.01$) and varsity sports ($r=-.22$, $p<.01$).

It is important to note that these correlations do not demonstrate causation; we cannot be sure if the participation in an activity increased related skills or if higher skills lead to participation in the related activity. However, this data indicates students' self concepts of what activities impacted their perceived abilities in desired on-the-job skills, which connects this research to a greater body of research exploring the role of student perceptions and self concept in preparing for job readiness.

Discussion

A primary goal of this study was to identify perceptions college undergraduate students hold about top skills desired in the workplace. National Association of College and Employers Job Outlook 2010 informed us that employers most heavily value communication skills, analytical skills, teamwork skills, technical skills, and work ethic, in that order. Students rated their work ethic and teamwork skills among their highest abilities, which is good news for the employers. Unfortunately, employers also seek analytical skills and technical skills, which students rated in the lower range of their abilities. These findings help identify a gap in the skills employers seek and students' perceptions of their abilities in those areas.

On a positive note, students were more in line with employers about perceptions of the importance of communication skills in the workplace. Perhaps initiatives at this specific university to infuse writing requirements into a wide variety of disciplines have convinced students that these skills are valuable. Ratings of the importance of communication skills did not differ between gender, class year, or academic discipline, supporting the possibility that a holistic academic approach to communication might be a factor in these results. On the other hand, this data was collected within a general education communications course, in which communications is obviously central to course success; if the data were collected in a different setting, these findings could dissipate.

It appears that academic involvement that requires what one might classify as active and on-going participation and effort on the part of the student (undergraduate research, internship/coops, and tutoring) are related to perceived desired skills. More passive types of academic involvement (faculty advising, receiving tutoring, and attending workshops) do not appear to have a relationship with key skills analyzed in this study. These results highlight the importance of active learning opportunities in which students construct new knowledge through the discovery of new ideas and participation in challenges that allow them to build upon previously gained skills (Bruner, 1960).

Also of interest was the relationship between undergraduate research and both analytic and major-specific skills. The university at which this study was conducted supports undergraduate participation in scholarly research with faculty mentors by providing both financial resources to faculty and academic credit to undergraduates. Though this program has been recognized as a national best practice in undergraduate research from a faculty perspective (O'Shaughnessy, 2008), this current study begins to fill the lack of research conducted to measure student perceptions of the program's value as related to students' futures. Overall, the data from this study demonstrates the effectiveness of such hands-on research at the undergraduate level, which can serve as encouragement for other universities to adopt similar programs.

Limitations

One of the primary limitations of this study is the correlational nature of the data. The survey is a cross-section of students' perceptions and does not follow changes in students' perceptions over time. In addition, we cannot make claims about the causal direction of these relationships. For example, the relationship between undergraduate research participation and analytical skills could exist because students who conducted research became more confident of their analytical skills. On the other hand, students who felt confident in their analytical skills might have decided to join undergraduate research teams. Even with this single-method bias, the relationships demonstrated here still provide valuable information about students' perceptions of their skills and the types of extracurricular and academic programs in which these students participate.

Because the data was collected within a communications course, the curriculum presumably included the importance of communication within the context of this course. Future research should determine if the results would be replicated in a different course, such as a general education mathematics course that would also include diversity of participants yet balance out any course-specific biases. While the participants here did represent a cross-section of colleges, years in school, and personal demographics, a larger sample size would support greater generalizablility. It is also important to remember that this study looked at the perceptions of traditional aged students in a four-year institution. Many other student populations exist and warrant their own investigations, including community college, non-traditional aged, minority (Teng, Morgan, & Anderson, 2001), and first-generation students (Moss, 2003).

Future Implications and Research

Considering that students graduating in 2009 faced up to 40 percent fewer job prospects than their counterparts faced the year before (Yousuf, 2009), the connection between college and career warrants continued study. For institutions struggling with attrition, the extra work needed to create and maintain programs and curriculum focused on career preparation might not be a top priority. Still, as students enter college seeking a career (Farrell, 2007), higher education cannot afford to ignore the issue of career preparation. In particular, the employer preference that appears to be unwavering no matter the economy or job market is the desire for "relevant work experience" (Koc & Koncz, 2009). Placing students in valuable, field-related work experiences (such as internships and co-operative learning placements) should be a high institutional goal.

Some educational critics, however, question such focus on career preparedness in college, stressing that focusing too heavily on career development comes at the expense of students' intellectual growth (Aronowitz, 2000; Moss, 2003). This view presents an institutional hurdle that must be cleared before campus-wide efforts can receive the financial and human resources needed to create and sustain initiatives for career preparation. In the end, however, the hope is that this research serves as a reminder to institutions of higher education to remember students' desire for career preparedness in their extracurricular and academic offerings, so these institutions can not only help students find jobs after graduation but to also prepare them—and help them see the value of career-related skills—so they might find long-term career success.

TABLE 1:

Significant mean difference in skills by class year

	Freshman	Sophomore	Junior	Senior
Perceptions of Analytical Skills				
Freshman	(3.73) [†]	-.22	-.49*	-.61*
Sophomore		(3.95)	-.26	-.38
Junior			(4.21)	-.12
Senior				(4.02)
Perceptions of Major Skills				
Freshman	(3.33)	-.47*	-.84*	-.94*
Sophomore		(3.80)	-.38	-.47
Junior			(4.18)	-.10
Senior				(4.28)
Perceptions of Teamwork Skills				
Freshman	(4.00)	-.49*	-.39*	-.50*
Sophomore		(4.49)	.09	-.01
Junior			(4.39)	-.11
Senior				(4.34)

* Significant at the p<.05 level.
† Mean scores are listed on the diagonal

TABLE 2:

Mean self-efficacy ratings of student skills compared to mean importance ratings from employers.

	Mean Student ratings of their skills	Mean Employer ratings of importance†	Mean Difference	t	df
Work ethic	**4.4**	4.6	-0.2*	-2.28	123
Friendly/Outgoing	4.4	3.7	0.7*	9.80	123
Teamwork	**4.4**	4.5	-0.1*	-2.29	123
Problem solving	4.1	4.5	-0.4*	-5.73	123
Flexibility/Adaptability	4.1	4.3	-0.2*	-2.92	122
Sense of humor	4.1	3.0	1.1*	18.99	123
Leadership	4.1	4.1	0.0	-0.37	122
Self-confidence	4.0	3.9	0.1	1.82	123
Initiative	4.0	4.5	-0.5*	-6.71	123
Organizational	3.9	4.0	-0.1	-0.69	123
Communication skills	**3.9**	4.7	-0.8*	-11.08	122
Tactfulness	3.9	3.8	0.1	1.88	123
Detail oriented	3.9	4.1	-0.2*	-2.06	123
Analytical skills	**3.8**	4.4	-0.6*	-7.53	123
Creativity	3.8	3.6	0.2	1.78	123
Strategic planning	3.8	3.3	0.5*	5.93	123
Computer skills	3.6	4.2	-0.6*	-6.55	123
Technical skills	**3.5**	4.1	-0.6*	-7.53	123
Entrepreneurial	3.3	3.2	0.1	1.79	123

* **The mean difference is significant at the 0.05 level.**
† Ratings of importance provided by employers in NACE Job Outlook 2010.

REFERENCES

American College Personnel Association. (1996). *The student learning imperative: Implications for student affairs.* (1996). [Electronic version]. Washington, DC: Authors.

Aronowitz, S. (2000). *The knowledge Factory: Dismantling the corporate university and creating true higher learning.* Boston: Beacon Press.

Bandura, A. (*1997*). *Self-efficacy: The exercise of control*. New York: Freeman.

Baxter Magolda, M. B., & King, P. M. (2004). *Learning partnerships: Theory and models of practice to educate for self-authorship.* Sterling, VA: Stylus Publishing.

Betz, N. E. & Hackett, G. (2006). Career self-efficacy theory: Back to the future. *Journal of Career Assessment, 14*(1), 3-11.

Bottner, R. (2010). Internship insights: A report from the national internship and co-op study. *NACE Journal, 70*(3), 26-28.

Bruner, J. (1960). *The Process of Education.* Cambridge: Harvard University Press.

Chickering, A. W. (1969). *Education and identity* (2nd ed.). San Francisco: Jossey-Bass.

Davidson, W. B., Beck, H. P., & Milligan, M. (2009). The college persistance questionnaire: Development and validation of an instrument that predicts student attrition. *Journal of College Student Development, 50(4)*, 373-390.

Dundes, L. & Marx, J. (2006/2007). Balancing work and academics in college: Why do students working 10 to 19 hours per week excel? *Journal of College Student Retention: Research, Theory & Practice, 8*(1), 107-120.

Farrell, E. F. (2007). Report says freshmen put career prep first. *Chronicle of Higher Education, 53*(18), A32.

Garber, J. (2003). *Getting a job.* New York: Barnes and Nobles Books.

Gundry, L. K., & Rousseau, D. M. (1994). Critical incidents in communicating culture to newcomers: The meaning is the message. *Human Relations, 47*, 1063-1088.

Hart, P. D. (2006). *How should colleges prepare students to succeed in today's global economy? Based on surveys among employers and recent college graduates.* (Association of American Colleges and Universities report). Retrieved from http://www.aacu.org/leap/documents/Re8097abcombined.pdf

Hannerman, L., & Gardner, P. (2010). Under the economic turmoil a skills gap simmers. Collegiate Employment Research Institute, Michigan State University.

Heckman, J. (1999). Education and job training: Doing it right. *Public Interest, Spring 1999*(135), 86-107.

Holland, J. L. (1997). *Making vocational choices* (3rd ed.). Odessa, FL: Psychological Assessment Resources.

Kim, M., & Alvarez, R. (1995). Women-only colleges: Some unanticipated consequences. *Journal of Higher Education, 66*(6), 641-668.

Koc, E. (2010, April 15). Hiring up 5.3 percent for class of 2010. *National Association of Colleges and Employers*. Retrieved from http://www.naceweb.org/Press/Releases/Hiring_Up_5_3_Percent_for_Class_of_2010_%284-15-10%29.aspx?referal=pressroom&menuid=273

Koc, E. & Koncz, A. (2009). *Job Outlook 2010*. Bethlehem, PA: National Association of Colleges and Employers.

Levine, A. L. & Cureton, J. S. (1998). *When hope and fear collide*. San Francisco: Jossey-Bass.

Louis, M. R. (1980). Surprise and sense making: What newcomers experience in entering unfamiliar organizational settings. ***Administrative Science Quarterly***, *25*(2), 226-251.

Moss, G. (2003). Intellectualism vs. career preparation: A comparative model of self reported growth among graduating college seniors. *College Student Journal, 37*(2), 309-318.

Morgeson, F. P., Campion, M. A., & Levashina, J. (2009). Why don't you just show me? Performance interviews for **skill**-based promotions. *International Journal of Selection & Assessment, 17*(2), 203-218.

O'Shaughnessy, L. (2008). *The College Solution: A Guide for Everyone Looking for the Right School at the Right Price*. Upper Saddle River, NJ: FT Press.

Pascarella, E. T., & *Terenzini*, P.T. (*2005*). *How college affects students (vol. 2): A third decade of research*. San Francisco: Jossey-Bass.

Petrecca, L. (2010, April). Grads' toughest test? Job market. *USA Today*. B1-B2.

Pittenger, K. S., Miller, M. C., & Allison, J. (2006). Can we succeed in teaching business students to write effectively? *Business Communication Quarterly, 69*(3), 257-263.

Pomerantz, N. K. (2006). Student engagement: A new paradigm for student affairs. *The College Student Affairs Journal, 25*(2), 176-185.

Rampell, C. & Hernandez, J. C. (2010, April 2). Signaling jobs recovery, payrolls surged in March. *New York Times*. Retrieved from http://www.nytimes.com/2010/04/03/business/economy/03jobs.html?scp=4&sq=job%20market&st=Search

Shaffer, J. P. (1995). Multiple Hypothesis Testing. *Annual Review of Psychology, 46*, 561–584.

Simon, A. (2010, May). Job outlook for college grads better than last year. *The Greenville News*, A1, A4.

Super, D. E. (1963). Self concepts in vocational development. In D. E. Super, R. Stariskevsky, N. Matlin, & J. P Jordaan (Eds.), *Career development: Self-concept theory* (pp. 1-26). New York: College Entrance Examination.

Sunshine, R. A. (2010, February). *The Budget and Economic Outlook: Fiscal Years 2010 to 2020.* Congressional Budget Office. Retrieved from http://cbo.gov/ftpdocs/108xx/doc10871/Chapter2.shtml#1105094

Tagg, J. (2003). *The learning paradigm college.* Bolton, MA: Anker.

Teng, L. Y., Morgan, G. A., & Anderson, S. K. (2001). Career development among ethnic and age groups of community college students. *Journal of Career Development, 28*(2), 115-127.

Tinto, V. (2006-2007). Research and practice of student retention: What next? *Journal of College Student Retention: Research, Theory & Practice, 8*(1), 1-19.

U.S. Bureau of Labor Statistics. (2010). *Occupational Outlook Handbook.* Washington, DC: U.S. Government Printing Office. Retrieved from http://www.bls.gov/OCO/

Walgran, K. (2010). Don't drunk-dial HR: Creating soft-skills workshops for college interns. *NACE Journal, 70*(3), 29-35.

Yousuf, H. (2009, November 18). *Job outlook for 2010 grads: Still stinks.* CNNMoney.com. Retrieved from http://news.msu.edu/story/7116/

Combining the Liberal and Useful Arts: Sociological Skills in the Global Economy

MARVIN S. FINKELSTEIN

The discipline of sociology remains vulnerable in an environment of economic uncertainty and global change. Constraints on higher education are likely to increase and recurrent pressures on traditional liberal arts programs will continue unabated. An older, more diverse, cost-conscious and career-minded student population will increasingly insist on clearer pathways to difficult and bewildering labor markets. But sociology's weakness as a liberal art may be overcome by combining it with a more applied and practical orientation. The very forces that threaten the discipline's institutional existence make it profoundly relevant and valuable in an age of social transformation. Based on a familiar Millsian conception of the sociological imagination, this article attempts to combine sociology's liberal tradition with its role as a "useful art," honed into the specific features of workplace change and the employment setting. It does so by suggesting five categories of emerging skills in the global economy and ways that sociology has a far reaching claim to their practice and development. The categories are: 1) the skills of knowledge workers; 2) skills in the learning organization; 3) skills in the technological context; 4) skills in the diverse and divided workplace; 5) change-making skills. The article concludes by urging those in the discipline to make sociology more of a useful art that has practical application in a changing world.

> *The transformation of this system from the earlier economy is so complete that it resembles more the metamorphosis of the caterpillar than any alteration that can be expressed in terms of continuous growth and development.*
> —*(Polanyi, 1957, p. 43.)*

Marvin S. Finkelstein is associate professor of sociology at Southern Illinois University at Edwardsville. Address for correspondence: Marvin S. Finkelstein, Department of Sociology and Social Work, Box 1455, Southern IL University, Edwardsville, IL 62026-1455.

> *What they need, and what they feel they need is a quality of mind that will help them to use information and to develop reason in order to achieve lucid summations of what is going on in the world. It is this quality of mind, ... what may be called the sociological imagination.*
> —(Mills, 1959, p. 5.)

Sociology is at a crossroads. It may continue on its present course emphasizing its role in academia, liberal arts, and basic research, or it may change direction toward a more applied, clinical and practical orientation. Of course, these are not mutually exclusive paths and some movement narrowing the distance between them is already occurring. However, it is clear that the latter of the two has been subordinated or blocked as a route leading to what the discipline has viewed as bonafide sociology. The change would likely consist of advancing sociology as an applied science and a "useful art," developing systematically an applied and clinically oriented curricula, encouraging research that is action-based and connected to addressing societal needs and concerns, and placing far greater emphasis on the way sociology prepares students for work and jobs in the twenty-first century. It is this last point in particular, that I give the greatest attention in this article.

The irony is that sociology is one of the few disciplines in the liberal arts or elsewhere with the potential to meet the challenge of higher education in a changing world. The emergence of a global economy—one which is far more integrated, competitive, dynamic and diverse than ever before, will require new kinds of skills and capabilities for the workforce. The challenge is to produce graduates who possess broader, deeper action-based skills, and a "quality of mind" that will help them solve complex problems creatively and cooperatively. To accomplish this task, sociologists will have to shift the discipline into a more applied and practical mode of knowing and learning. The goal of this article is to 1) underline the need for sociology to combine its liberal tradition with the practical orientation of the "useful arts," 2) draw a closer connection between sociology and emerging skills in a global economy 3) focus on five categories of skills and the ways sociology possesses a far reaching claim to their practice and development.

Trends and Developments

Discussion within the discipline about the future of sociology in the United States has intensified in the last few years (Freeman and Rossi, 1984; Rossi, 1988; Gollin, 1990; Lynch, Mcferron, Bowker and Bedford, 1993; Brown, forthcoming). Although it appears that enrollments and majors have held steady of late and that pendulum-swinging students are on the way back to the social science fold, the recognition of the discipline's vulnerability continues to grow in the wider context of increasingly constrained resources. Consider these developments and trends:

- State support for public higher education has slowed to a crawl compared to the 1980s and is not likely to recover anytime soon, if at all (Lively and Mercer, 1994; Lively, 1993). Private liberal arts colleges will continue to face similar pressures and constraints (Breneman, 1994).
- Deep budget cuts and tuition increases may cause enrollments to drop and competition for college-bound students to increase in the coming years, despite projected increases in the pool of high school graduates. Besides California, which has been hit the hardest, thirteen other states have reported enrollment decreases for similar reasons during the 1992-1993 academic year (Lively 1994; Hohm, 1992).
- Demographic trends indicate an older and more diverse college-student population interested in fulfilling practical objectives. By the year 2000 over half of all college students will be over the age of twenty-five (Simpson and Frost, 1993; Gallos, 1991).
- The number of students who attend college primarily to prepare for a career and improve their economic fortunes is increasing. Seventy-nine percent of entering college freshmen in 1991 reported that getting a better job was "a very important" reason in their decision to attend college, up from seventy-one percent in 1976. Moreover, in 1976, fifty-four percent of entering freshmen said that "making more money" was a very important reason for attending college—in 1991 the number was seventy-five percent (Astin, 1993).
- College freshmen appear to be far more interested in careers in business, law and medicine and are much less interested in college teaching or research careers than they were 20 years ago (Useem, 1989; Astin, 1993).
- College graduates will continue to face an increasingly difficult and restrictive labor market and graduates in the humanities and social sciences will be the most seriously affected (Sharp and Weidman 1989). The numbers of college graduates taking jobs that don't require a college degree are rapidly rising. Almost a third of new college graduates will be underutilized between now and 2005 (Church, 1993).
- Employers are now spending more money on postsecondary education *outside* of universities and colleges than is spent on the entire system of higher education in the United States (Carnevale, 1989).

These trends describe a very different world than the one in which sociology grew and flourished a generation ago. They suggest that in the future, fiscal constraints will lead American higher education to further consolidate resources as cost-conscious and career-minded students sharpen their focus on academic programs that will help them maneuver in a chronically restrictive labor market. One clear implication for the discipline may be contraction: as retirements and attrition increase, unsympathetic deans and administrators may siphon off faculty lines and new positions to other areas (Lynch, Mcferron, Bowker and Beckford, 1993).

But these trends must be placed in the wider context of unprecedented global change. Organizations of all kinds, academic and nonacademic alike, are experiencing these pressures and constraints. Universities and colleges are facing "post industrial environments," in which lean resources, skyrocketing tuition, declining enrollments, competitiveness, turbulence and uncertainty increasingly characterize the terrain of higher education (Cameron and Tishart, 1992). The post-World War II boom of economic activity in America that supported and helped shape the massive growth of post secondary institutions has changed forever.

In that era, liberal education could comfortably distance itself from practical concerns and remain committed to purely academic goals: "knowledge for the sake of knowledge." "Established" sociology, encouraged the role of "disinterested" research to help raise the status of the discipline as legitimate science. The challenge by conflict sociology only reinforced this position by making any connection to business and industry anathema. Baby boomers, born and bred on infinite expectations for social justice and equality in a booming economy, flocked to sociology with questions and interests reflected in those ideals. But as sociology fought its internal ideological and curricular battles with great conviction and courage in the 1970s, global change continued to accelerate as did the decline of U.S. economic dominance.

Today, students still look toward the liberal arts for intellectual growth and moral guidance, but they have come to demand clear practical pathways toward becoming engaged in the world. These pathways, however, will be difficult to discern in an economy and labor market that will likely remain ambiguous and uncertain. Entire industries, occupations and careers will ride a roller coaster of change. Employers increasingly will stress the broad capacities obtained in the liberal arts, but they will become far more interested in those who have grounded their liberal learning in skills that are honed into the practical features of an employment setting. Even if these needs are met, it appears that state support for higher education will continue to shrink while private-sector training continues to grow. It is quite possible that nonacademic training could become the dominant form of higher education in the United States. Moreover, without solid roots in the community, industry, and the public sector, traditional sociology will lack public recognition of its relevancy and effectiveness. In brief, the conditions that permitted sociology and other liberal arts disciplines to thrive in the post-World War II era have collapsed. The question is whether sociology is making the changes necessary to meet the challenge. There is reason to believe the answer is far too little.

One recent important proposal for change has, for example, come from efforts by the American Sociological Association urging sociology departments to promote study in depth by restructuring the curriculum and sequencing courses (Sociology Task Force, 1990). This proposal is an important first step, however, it does little to reframe traditional liberal arts assumptions and continues to define the discipline as almost exclusively a "scholarly endeavor" providing "intellectual development." Indeed, of the thirteen recommendations made by the Sociology Task Force, just one of them mentions "application" and does so with regard only to policy issues. While proposals to restructure curricula toward systematically attainable goals are well taken, these lack sufficient attention to bridging the gap between traditional liberal arts and what Ernest Boyer has termed "the useful arts" (Useem, 1989). In the latter, scholarly endeavor and intellectual development are imperative, but they are thoroughly immersed in practical matters of interest to employers and students alike (Boyer, 1987).

Accomplishing this task would require a reconciliation and synthesis of the liberal arts and practice traditions (Green, 1993). It would mean clearly developing the content and relevancy of curricula in relation to emerging workplace issues and skills. As it stands, there is a great deal of confusion as to what applied sociology or sociological practice is, its professional identity in the job market, and how it should be presented to students in the classroom (Ruggiero and Weston, 1991). This is especially significant at the graduate level where few clinical masters programs can be found nationwide and where Ph.D. programs consistently lack a clear connection to sociological practice or sociological practice associations (Fritz and Clark, 1993). In addition, sociology as a useful art must translate into creating stronger ties and cooperative partnerships between the state, industry and education (Johnstone, 1994; Whyte, 1991). Without graduates who have found and developed professional careers in community and who recognize their educational roots as useful and relevant, public support for sociology will further erode and the discipline's future will remain uncertain (Fleischer, forthcoming).

Global Change and Sociology

The discipline of sociology was born almost two centuries ago in the context of a societal sea change, what Polanyi (1957) called the "great transformation." Much of sociology may be understood as an effort to comprehend and explain the significance of this momentous change. Today, there is evidence to suggest that a change of equal or greater proportions may be underway (Bell, 1976; Toffler, 1980; Harvey, 1992). This is particularly true with regard to the nature of production, work and employment (Reich, 1983; Sabel and Piore 1984; Womack, et. al, 1990). In general, the change has to do with the movement away from the "Fordist" model of organization characterized by a high degree of bureaucracy, Taylorism, mass-production techniques and products, toward "lean" or "flexible" systems featuring flatter, more highly integrated, team-based operations fueled by rapid information flows and geared for innovation and service-oriented products. Few areas of social life are left untouched by these changes. They encompass the impact of new technologies, demographic shifts in the workforce, work and family relationships, gender, race and class issues as well as education, training and public policy. We are confronted with a world increasingly "compressed" in time and space, economically integrated and interdependent, and in a state of permanent flux.

Despite its sociological content, most of the discussion focusing on workplace and industrial change has come from the flood of business books, consultants and from other fields such as management, industrial psychology, organizational behavior and development, (which have not hesitated to draw substantially from sociology). Sociologists, on the other hand, have not only distanced themselves from practical involvement in workplace and industrial change, but they have virtually ignored the implications of these vast changes for the future of the discipline (Finkelstein, 1990, 1993).

Reasons for this distanced approach again may be traced back to objections from either side of the political spectrum: those who view such activity as too great a compromise of the "interest free" imperatives of scientific protocol on one end, and those who see practical involvement as assisting the powerful in their attempts to exploit and manipulate the less powerful, on the other. Yet these views continue to become increasingly blurred as action-based, participatory methodologies take on greater legitimacy in the scientific community (Park, 1994; Whyte, 1991), and more attention is focused on the possible convergence of traditional business goals such as efficiency, productivity and profitability, with the goals of worker dignity, equality, and democracy so strongly held by advocates of social change (Bluestone and Bluestone, 1993; Kuttner, 1991; Block, 1990).

Workplace change increasingly deals with various forms of employee participation in decision making, information sharing and economic ownership and reward systems. It also involves an incredibly complex and conflictual struggle to create the conditions necessary to accomplish these goals. More often than not, success or failure in private and public organizations alike depend on the degree of commitment to change and the requisite skills to create change. In short, it is important to recognize that sociological practice in the workplace may be both scientific and humanistic (Fritz and Clark, 1989), and that it offers our students the opportunity of applying what they know not only to meet their career aspirations but to foster social improvement and change (Rossi and Whyte, 1983).

The Need for a Quality of Mind

There is little doubt that "the old ways of doing things" is being called into question in virtually every aspect of the world of work. However, redefining the role of management, the role of labor, hierarchical relations, or doing away with gender/racial stereotyping and segregation along with a myriad of other workplace relics, will take extraordinary imagination and commitment. To create what many observers view as tantamount to a "revolution" in a sphere where straight-jacketing work roles, fragmentation, conflict, and distrust have deep intractable roots, will require new ways of thinking, learning and doing. These skills are not reducible to technical knowledge. While computer, statistical and methodological competencies are extremely important in the age of the electronic text, they cannot be narrowly defined and they must cut across a variety of functional areas. Specialized skills, which appear relevant today may be obsolete tomorrow, thus requiring skills that are sufficiently technical to do a job but broad enough to adapt and make changes as needed (Gordon, 1991). Moreover, the expansive capacity to make change and innovate may be one of the most important skills to be learned in an environment in which change is the only constant.

In fact, many are calling this the age of the "specialized generalist" stressing the need to "generalize the workforce" so that there are more of those who can perform many different jobs, cut across a variety of functional areas and pull together with others to coordinate activities and projects (Kaeter, 1993). When one considers, for example, that the ideologies of both individualism *and* "groupism" continue to distort our understanding of effective small group activity and hinders our ability to establish participatory forms of organization and decision making, then the kind of knowledge necessary to encourage change becomes clearer and more tangible (Cole, 1989).

Technical skills must be combined with conceptual, critical analytical and small group skills within the framework of a cooperative problem-solving process. They must be steeped in a broad historical and scientific understanding of human behavior and in the importance of social context, the collectivity and the environment in the shaping of behavior. They must also be skills in which practice is always a reflective process rather than a mechanical one. These are expansive skills, not vocational ones.

In sociology, there has been a valiant effort to move beyond the insipid reference to "people skills," and to provide an in-depth description of skill categories (Ballentine, 1989; Brown; forthcoming). But to demonstrate the connection between sociological skills and global change, it may be useful for our purposes to focus on one capacity that underlies many if not all of them; one that you might find surprisingly familiar: *the ability to understand, challenge, critique taken for granted assumptions, perspectives or worldviews that often trap us in old categories or ways of thinking and deny us the potential for transforming our circumstances creatively.* The notion that we are often trapped in prisons of our own making—cultural beliefs, attitudes, accepted practices, and that we may free ourselves by becoming aware of the influence of social forces on our lives, is of course something familiar to all sociologists.

You would probably have little trouble finding such a definition or statement in the first chapter of almost every introductory sociology textbook. Call it what you will—C.W. Mills's the "sociological imagination," or Berger's "invitation to sociology," or merely the sociological perspective, the contention here is that meeting the challenge of global change will require a "quality of mind" capable of understanding and challenging straight-jacketing assumptions of the old ways of organizing and doing work, to create new social relationships that free employees to be more creative and innovative on the job. A sociology that reorients its liberal-arts tradition toward such application and practice has far-reaching implications for emerging workplace skills in a rapidly changing global economy.

The literature in the area of changing work, organizations and industry is incredibly vast and unwieldy. Much of it, as previously suggested, has been done by nonsociologists even though they may have drawn heavily from sociology. There are, however, recurrent themes and motifs when it comes to addressing the kinds of skills needed by managerial-professionals to foster workplace change or to work in environments where change, flexibility, cooperation and small-group activity are commonplace. The last section of this paper attempts to establish a clearer connection between recent insights on emerging skills identified in this literature and the "quality of mind" cultivated in the sociological perspective. This may be accomplished by examining five categories of skills, which may help us better understand sociological practice in the global economy: 1) skills of knowledge workers, 2) skills in the learning organization, 3) skills in the technological context, 4) skills in the diverse and divided workplace, 5) change-making skills. Taken together, these sociological skills may produce a better understanding of how the liberal and useful arts may be combined in the employment setting.

Sociological Skills in the Global Economy

The Skills of Knowledge Workers: Symbolic Analysts

One of the fastest and largest-growing segments of the workforce is professional and technical workers who, by the year 2005, may compose 20% of all workers or over 23 million people (Kiechell, 1993; U.S. Dept. Of Labor, BLS, 1993). As technical knowledge proliferates and more technically oriented workers demand or require the conditions of professionalized work, such as greater autonomy and control, this segment will continue to grow (Hall, 1994). Add to this the category of managerial and administrative employees and the percentage approaches about a third of the workforce who constitute what sociologists have for a long time viewed as a "new class" (Bell, 1976; Bruce-Briggs, 1979; Gouldner, 1979). Central to this new class is a knowledge base that provides members with abstract, specialized expertise and the society with a powerfully new productive force—information.

The "new class" idea has been renewed in Robert Reich's, *The Work of Nations* (1992), which builds its thesis around the contention that for America to regain its economic competitiveness in the global economy, it will have to encourage the growth and development of a category of workers he calls "symbolic analysts." According to Reich, symbolic analysts have the knowledge and skill to add value to products that the world wants, as opposed to other categories of workers who merely do routine production or service jobs. Symbolic analysts are distinguished by four skills: abstraction, systems thinking, experimentation and collaboration. Essentially, these are higher order skills that provide workers with the ability to be problem-identifiers, problems solvers and information brokers.

With the capacity for abstraction, for example, Reich argues that the focus is on judgment and interpretation, and not just the transmission of information. "The student is taught to get *behind* (author's emphasis) the data—to ask why certain facts have been selected, why they are assumed to be important, how they were deduced, and how they might be contradicted. The student learns to examine reality from many angles, different lights and thus to visualize new possibilities and choices" (Reich, 1992:230). To go one step further, Reich claims that systems thinking, requires the capability of "seeing the whole, and of understanding the processes by which parts of reality are linked together," trying to discover, "larger causes, consequences and relationships" (Reich, 1992:231). Experimentation means that "students are equipped with a set of tools to find their own way. The focus is on experimental techniques: holding certain parts constant while varying others to better understand causes and consequences; systematically exploring a range of possibilities and outcomes noting similarities and differences; making thoughtful guesses and intuitive leaps and then testing them against previous assumptions" (Reich, 1992:232). Finally, collaboration is preparing for "lifetimes of symbolic-analytic teamwork," small group learning, consensus and conflict resolution (Reich, 1992:233).

Do these sound like familiar objectives that are crucial to introductory sociology, theory or research methods and statistics courses? They clearly parallel the "quality of mind" outlined by C.W. Mills over thirty years ago. Thus, we find that Reich, the current U.S. Secretary of Labor, is identifying skills that are largely sociological, especially when taken together—uncovering and critically assessing the "facts," assumptions, finding contradictions, relating individual experiences or parts of a problem to the whole of society or an organization, doing multivariate analysis, understanding groups, organizations and conflict—and he is contending that for the United States to avoid further economic decline, class and income polarization, it must invest in expanding the numbers of symbolic analysts who hold the key to twenty-first century competitiveness in their ability to lead the world in knowledge creation. Sociology students should know that what they are being taught offers invaluable practical application in virtually any career they might choose.

Skills in the Learning Organization: Learning Skills

The sociological skills of symbolic analysts reappear in efforts to develop organizations in which employees of this kind might flourish. But to create and foster new kinds of employees requires new kinds of organization. One of the most important recent developments in the workplace change literature is the concept of the "learning organization." A recent special issue of the influential journal *Organizational Dynamics* dedicated to this topic, in the words of the editors, fulfills its mission to "link leading-edge theory and research in the behavioral sciences with management practice" (Luthans and Bohl, 1993).

Building especially on the "action based" social psychological theories of Argyris and Schon (1974, 1978), Peter Senge (1990), Director of the Systems Thinking and Organizational Learning Program at MIT, has become one of the leading proponents of skills, which provide organizational members with the tools necessary for continuous and collective learning. Senge claims that there are five "technologies" that comprise the making of a learning organization: 1) systems thinking 2) personal mastery, 3) mental models 4) building shared vision and 5) team learning. Like Reich (1992), Senge focuses on systems thinking, which is what he calls "the fifth discipline" since it integrates the other four into a new way of thinking and acting. He argues that the fifth discipline provides skills, which, above all, allow learners to see the world as interrelated parts of the whole, instead of narrow, isolated snapshots.

With little or no mention of a sociological perspective (except one reference to Robert Merton—as a psychologist!), Senge underlines other important, though rather elementary sociological lessons such as "structure influences behavior" and "structure in human systems is subtle." Moreover, he warns against organizational "learning disabilities," such as thinking that is excessively individualistic, the failure to recognize how one's own actions contribute to the making of the whole and the fragmentation of social life so characteristic of bureaucracy. Echoing the "quality of mind" mentioned earlier, Senge defines the fifth discipline as "a shift in mind—from seeing ourselves as separate from the world to connected to the world, from seeing problems as caused by someone or something 'out there' to seeing how our own actions create the problems we experience. A learning organization is a place where people are continually learning how they create their reality. And how they can change it." Thus, the fifth discipline focuses on group feedback, "double loop learning" and assessment to help organizational members solve problems and make changes collectively and effectively.

Though these may appear to be rather rudimentary sociological points applied to the organizational context, their value should not be underestimated. The problem is that students of sociology do not often see the potential force of their own learning because it is not framed as a skill now recognized, for example, as "leading edge" in the field of management. By reframing basic sociological concepts in the context of global economic change, we may make available to our students a liberal art made useful, the art of learning to learn, which is becoming central to any productive enterprise.

Skills in the Technological Context: Intellective Skills

Symbolic analysts operating in a learning organization will be confronting new technology that will intensify the need for new kinds of skills. Blauner (1964) indicated as much when he presented his well-known inverted U curve of alienation at work. There he contended that settings, which move away from mass production techniques toward more advanced continuous process technologies, require higher levels of skill and thus less alienation. Of course, others take the opposite position that skills will deteriorate in the face of tightly controlled, top-down technical control (Braverman, 1975; Edwards, 1979). This is the stuff of sociological debates, but they should not be viewed as purely academic. As we have argued, they have real practical value for our students and the community.

One of the most cogent analyses of technological change is presented by Shoshana Zuboff (1988), a social psychologist, who advances further, not only in understanding such debates, but in identifying the types of skills that are emerging as computer technology becomes incorporated into the production process. For Zuboff, this is nothing short of a revolution at work in which all types of production, thought and action become abstracted and expressed on computer screens in a new electronic language. Thus, work is no longer a purely physical-sensual labor but rather always computer-mediated. Whatever is produced, whether goods or services, is transformed as feedback in a computer system. Does the system enhance and foster the skills of workers or does it control, manipulate and deskill them?

This technological context provides the possibility for both scenarios. Zuboff develops the progressive alternative by suggesting the emergence of "intellective skills." They are remarkably akin to the ones suggested by Reich and Senge. Since all work is restructured into a symbolic medium, skills are based on abstract, sensemaking, and inferential reasoning—both deductive and inductive. According to Zuboff (1988:96), they involve ways of ascertaining "the condition of 'reality' in ways that cannot be reduced to correspondence with physical objects (for example, the ability to discern states, trends, underlying causes, relations, dynamics, predictions, or opportunities for improvement)." This involves troubleshooting and working through problems involving symbolic meaning, which always must be interpreted, reconstructed, reflected on and worked through, not in an isolated individual way, but in groups. The group provides a rich variety of perspectives that may be brainstormed, torn apart and put back together again in innovative forms.

Thus, for Zuboff, the proper interpretation of data as they appear on the computer screen is rarely self-evident. Interpretations are actively constructed to confer meaning on the data in a dialogue and as part of joint hypothesis testing. For symbolic analyzers, the real question of skill in the computer age is not so much "how do you do it?" or "what do you do?" It's "what does it mean?" Something that again calls for that quintessential sociological "quality of mind." Zuboff argues that given this "infomated" technological context, it is possible to provide instantaneous information access to large numbers of people in the organization whose possibilities for active participation and involvement in decision making become enormously expanded. As she suggests and Senge would agree, "(l)earning is the new form of labor," grounded in a new set of "intellective" activities. It is an environment where change, flexibility, cooperation and small group activity are commonplace.

Sociological skills in the technological context, therefore, can first, encompass the significance and implications of computers in the workplace by defining (de)skilling issues, second, inform those who are shaping the workplace of an alternative environment with these infomated learning possibilities, and finally, provide the capacity to deal with computers more effectively on the job.

Skills in the Diverse and Divided Workplace: Power Tools

The computer age appears to provide symbolic analysts who have a combined liberal and useful arts background in sociology, great potential for learning and development in the global economy. However these opportunities will likely continue to be scarce and unequally distributed. Perhaps more than any other kind of skill mentioned, the capacity to deal with stratification by race, gender and class must be viewed as a central strength of the sociological perspective (Baca Zinn and Eitzen, 1994), and must be made more usefully grounded if it is going to give sociology graduates the ability to have an impact on the world of work.

The demographic revolution of the next decade has become increasingly well known. Women will account for over half of all new entrants to the labor force and constitute nearly half of the labor force by the year 2005. In fact, 4 out of 5 women between the ages of 25 and 34 will be in the labor force. By 2005, African Americans, Hispanics, Asians and other minority groups will account for over a third of new labor force entrants and will represent 27% of the workforce. At the same time, as the baby "bust" takes hold, there will be a stronger demand for younger, highly educated and skilled workers, and diversity in the workplace will become an economic necessity as well as a moral obligation (Fernandez, 1991).

Moreover, Reich (1992) suggests that those not comprising the category of symbolic analyst fall into two other categories of work: 1) routine production services, which may or may not be supervisory but are largely concerned with bureaucratized tasks and functions that either remain in the dwindling mass production industries or are relegated to the low end of "hi-tech" production, such as data processing. Though relatively high paying, jobs in both these areas are on the decline. 2) In-person services may also be repetitive and routine, but they involve one-on-one contact with the customer or client. These include hotel, restaurant, taxi, hairdresser, security guards, orderlies, aides and attendants. This has been one of the fastest growing sectors of the labor market and as one might expect, growing numbers of women and minorities are relegated to this category of work. Thus, the persistent institutional problems associated with bureaucracy, occupational segregation, discrimination, glass ceilings and comparable worth continue to divide and debilitate the workforce and sap the productive capacity of all organizations.

There are few scholars or practitioners who have dealt with these issues more cogently than sociologist, Rosabeth Moss Kanter. Long before "diversity" and "multiculturalism" became hot topics in corporate America, Kanter (1977) was not only researching these problems but addressing them in practical ways. She gives us greater insight into the ways sociological skills can be brought to bear on these seemingly intractable problems. First of all, Kanter's analysis is primarily structural. Her first premise is "jobs create people." Behavior is guided by stereotypical gender roles and images that coincide with predetermined job classifications in the hierarchy. She unmasks a social system in which people maintain positions of power or powerlessness and play commensurate roles without often understanding where they came from or their negative effects. She instructs us to call into question the taken for granted corporate bureaucratic order and to recognize that, alas, the emperor has no clothes.

For Kanter, structural problems require structural solutions. Her prescriptions to "flatten the hierarchy," to decentralize operations, and balance the numbers among majority/minority groups, has become commonplace in the literature. For Kanter, the problems of bureaucracy—centralized control, fragmentation, alienation and ineffectiveness—are bound up with the problems of race and gender discrimination. At the social psychological level, she focuses on the micro skills needed to deal with such obstacles which leads us to her other well known aphorism, "empowerment."

Structural reorganization to foster team-based participation, for Kanter (1983), makes power more available to those who have been traditionally excluded by broadening the base of opportunities to become more involved with the decision making that affects their jobs. Empowerment means developing "power skills," that are essentially effective forms of persuasion and communication used to obtain the "power tools"—information, resources and support necessary to solve problems and get things done. Power in such a structural arrangement relies less on position, gender or race and more on the ability to persuade and convince others. Thus, very much like Reich, Senge and Zuboff, Kanter recognizes that skills that empower are those that harness knowledge, concepts and information. Kanter (1989) has continued to emphasize changing relationships and social processes at work—team building, coalition building and collaboration, necessary to mobilize activity. Taken together, skills steeped in creating structural change and empowerment amount to a powerful guide to sociological practice. They are skills sorely needed by our students and the organizations they work in.

Change-Making Skills: Imaginization

If there is one defining feature of global change, it is change itself. There is, of course, nothing new about social change. What is different today, however, is its accelerating rate, scope and magnitude. Moreover, because change is so rapid and unwieldy, social forces may move in many different directions at once. Thus, change is often contradictory, paradoxical, and underlying what is immediately "given" or perceived in everyday life. In short, there is a growing need for those who can make sense of the incredibly complex and multilevel nature of such changes, which otherwise might easily go unnoticed or appear totally mysterious and incomprehensible.

Sociology is uniquely qualified to fill this need (Finkelstein, 1992). There are few disciplines that are equipped with a perspective that can, for example, address simultaneously individual, group, organizational, cultural, societal, global and historical levels of analysis. But even more important than this is the capacity to provide a change perspective for those who wish to create change or who are engaged in change. One of the greatest obstacles to creating workplace change, outside of the tyranny of an entrenched elite, is simply a lack of inclination to be proactive and to be an agent of change. This lack of inclination may be traced back to the inability to conceive of alternatives and to think creatively. Indeed, recent research suggests that the latest frenzy to implement TQM (Total Quality Management) may fail and become another quick-fix gimmick, because creative change-making skills are lacking. The core of these skills has to do with the ability to defer judgment, to produce the widest variety of options possible and take deliberate action based on those considerations (Basadur and Robinson 1993). These skills have deep roots in sociology's liberal arts tradition, and they are key elements in what others have called "problem solving sociology" (Kapusinski, 1989).

But here we've gone full circle. The "quality of mind" suggested in Mills's sociological imagination is precisely that capacity to challenge existing arrangements, see things differently, to imagine new possibilities that may help liberate us from the chains of our own thinking. Perhaps the best example of the application of this approach and one of the most theoretically informed guides to practice in the field of workplace change is the work of Gareth Morgan. Morgan (with Burrell, 1979) distinguished himself as a major organizational theorist over a decade ago. But since then he has become increasingly involved with application and practice as a management consultant. Interestingly enough, though without referring to Mills, Morgan calls his approach "imaginization," which he defined recently as

> .... a way of thinking. Its a way of organizing. It's a key managerial skill. It provides a way of helping people understand and develop their creative potential. It offers a means to finding innovative solutions to difficult problems. And, last but not least, it provides a means of empowering people to trust themselves and find new roles in a world characterized by flux and change (xxix, 1993).

Imaginization builds on Morgan's (1986) efforts to propose metaphors and images, which may be used to challenge taken-for-granted assumptions, ideas and concepts and to demonstrate a variety of ways to "see" or "read" organizations of all kinds. Thus, he suggests that, for example, the "machine" metaphor has been a powerful image, which has guided both theories of organizations and behavior inside them. By utilizing skills rooted in his imaginization approach, Morgan encourages us to recognize a spectrum of images and metaphors—organizations as organisms, cultures, brains, political systems, instruments of domination—which help us to break with such mechanical categories and concepts that constrain our ability to "see" so that we may change. Morgan, drawing on the social constructionist tradition so strongly represented in sociology, provides the penetrating insight that the workplace is a prison of our own making. We create that prison everyday as we experience it. But recognizing the possibility of alternatives is central to the skills necessary to change it. This is a sociological skill, one that combines the liberal and useful arts.

Conclusion

Each of us has our own special relationship with sociology. What sociology is to one person may be quite different to another. One of the things we hold in common, however, is a distinctive approach, a "quality of mind" echoed by Mills and others, to understanding and explaining the world. Today the forces that gave birth to sociology are erupting in ways unparalleled in the modern age. These forces bring forth both peril and promise. At the same time the discipline appears vulnerable and in trouble, its perspective has become profoundly relevant and valuable. But to realize this potential will require combining the liberal arts and practice traditions. The challenge will be to reconceptualize that quality of mind as a "useful art." By doing so, the discipline may be more likely to overcome deep-seeded aversions to sociological practice and it will have a strong claim and contribution to new skills in a burgeoning global economy.

Students of sociology who can make that claim as symbolic analysts possessing learning skills and intellective skills, who know how to obtain power skills and know how to use them, and who have mastered the art of making change, will be better prepared for work in the global economy. These are skills that provide students of sociology with the tools to take a scientific approach to problem solving while at the same time the capacity to critically analyze the interests served by the solutions. What other discipline is equipped for such a task?

But if these skills are going to be available to our students and those interested in the discipline, sociology must be viewed more as a useful art that has practical application in a changing world. It must be presented systematically in curricula that reflect these qualities and it must begin to produce practitioners in the community who identify with its goals and support its existence. Our students don't deserve anything less. Our future may depend on it.

Note

An earlier version of this article was presented at the Sociological Practice Association 15th Annual Meeting, Denver, Colorado, June 11, 1993. I wish to thank Karen Olson, Bill Brown and Bradley Fisher for their encouraging support and helpful comments in preparing this manuscript.

References

Astin, Alexander W. 1993. *What Matters in College: Four Critical Years Revisited*. San Francisco: Jossey-Bass.
Argyris, C. and D. Schon. 1978. *Organizational Learning: A Theory of Action Perspective*. Reading, MA: Addison Wesley.
Argyris, C. and D. Schon. 1974. *Theory in Practice*. Reading, MA: Addison Wesley.
Baca Zinn, M. and D. Stanley Eitzen. 1993. "The Demographic Transformation and the Sociological Enterprise," *The American Sociologist*, 24, 2: 5–12.
Ballentine, J. 1989. "Developing Applied Content in Sociology Courses." *Journal of Applied Sociology*, 6: 89-94.
Basadur, M. and S. Robinson. 1993. "The New Creative Thinking Skills Needed for Total Quality Management to Become Fact, Not Just Philosophy." *American Behavioral Scientist*, 1: 121-138.
Bell, D. 1973. *The Coming of the Post-Industrial Society*. New York: Basic.
Blauner, R. 1964. *Alienation and Freedom*. Chicago: University of Chicago Press.
Block, F. 1990. *Postindustrial Possibilities; A Critique of Economic Discourse*. Berkeley: University of California Press.
Bluestone, B. and Irving Bluestone. 1992. *Negotiating the Future: A Labor Perspective on American Business*. New York: Basic.
Boyer, E.L. 1987. *College: The Undergraduate Experience in America*. New York: Harper and Row.
Braverman, H. 1975. *Labor and Monopoly Capital*. New York: Monthly Review Press.
Breneman, D.W. 1994. *Liberal Arts Colleges: Thriving, Surviving, or Endangered?* Washington, DC: The Brookings Institution.
Brown, William R. 1993 . "A Proposal Multi-Level Plan to Market Sociological Competencies." *The American Sociologist*, 24, 3 & 4: 87-105.

Bruce-Briggs, B. (ed.). 1979. *The New Class?* New Brunswick, NJ: Transaction Books.
Burrell, G. and G. Morgan. 1979. *Sociological Paradigms and Organizational Analysis*. London: Hieneman.
Cameron, K. and Mary Tschirhart. 1992. "Postindustrial Environments and Organizational Effectiveness in Colleges and Universities." *Journal of Higher Education*, 1: 87-108.
Carnevale. A. 1989. "The Learning Enterprise." *Training and Development Journal*, 46: 26-33.
Church, G.J. 1993. "Jobs in the Age of Insecurity." *Time*, (November 22): 34-39.
Cole, R. 1989. *Strategies for Learning: Small Group Activities in American, Japanese, and Swedish Industry*. Berkeley: University of California Press.
Edwards, R. 1979. *Contested Terrain*. New York: Basic Books.
Fernandez, J.P. 1991. *Managing a Diverse Workforce*. Lexington, MA: Lexington.
Finkelstein, M. 1992. "Taking Back a Rich Tradition: A Sociological Approach to Workplace and Industrial Change in the Global Economy." *Clinical Sociology Review*, 10: 182-197.
Finkelstein, M. 1990. "Sociologists Needed: But Will They Come Forward in America's Industrial Transformation?" *Sociological Practice Review*, 2: 71-76.
Fleischer, M.S. (Forthcoming). *State Licensing Monitors' Handbook*. Washington, DC: American Sociological Association.
Freeman, H. and P.H. Rossi. 1984. "Furthering the Applied Side of Sociology." *American Sociological Review*, 49: 571-580.
Fritz, J. and E. Clark. 1993. "An Overview of the Field of Sociological Practice: The Development of Clinical Sociology and Applied Sociology," in *Teaching Sociological Practice: A Resource Book*, C.B. Howery, N. Perrin, J. Seem and R. Bendiksen (eds.). Washington, DC: American Sociological Association.
Fritz, J. and E. Clark (Eds.). 1989. *Sociological Practice: The Development of Clinical and Applied Sociology*. East Lansing, MI: Michigan State University Press.
Gallos, J. 1992. "Educating Women and Men in the 21st Century: Gender, Diversity, Leadership Opportunities." *The Journal of Continuing Education*, (Winter): 2-8.
Gollin, A. 1990. "Whither the Profession of Sociology?" *The American Sociologist*, 21, 4: 316-320.
Gordon, J. 1991. "The Skilling of America." *Training*, (March): 27-35.
Gouldner, A. 1979. *The Future of the Intellectuals and the Rise of the New Class*. New York: Seabury Press.
Green, C. S. 1993. "Teaching Sociological Practice within A Liberal Arts Education," in *Teaching Sociological Practice: A Resourcebook*. C. Howery et al. (eds.). Washington DC: American Sociological Association.
Hall, R. 1994. *Sociology of Work: Perspectives, Analyses and Issues*. Thousand Oaks, CA: Pine Forge Press.
Harvey, D. 1989. *The Condition of Postmodernity: An Inquiry into the Origins of Cultural Change*. Cambridge, MA: Blackwell.
Hohm, Charles. 1992. "Some Reflections on Sociology's Future in California." *Footnotes*, (February): 5. Washington, DC: American Sociological Association.
Johnstone, D.B. 1994. "College at Work: Partnerships and the Rebuilding of American Competence." *Journal of Higher Education*, 2:168-182.
Kaetner, M. 1993. "The Age of the Specialized Generalist." *Training*, December, 48-53.
Kanter, R.M. 1989. *When Giants Learn to Dance*. New York: Simon and Schuster.
Kanter, R.M. 1983. *The Change Masters*. New York: Simon and Schuster.
Kanter, R.M. 1977. *Men and Women of the Corporation*. New York: Basic Books.
Kapusinkski, A. 1989. "Problem Solving Sociology: Learning Creative Problems Solving in an Undergraduate Seminar." *Clinical Sociology Review*, 7: 178-197.
Kiechell III, W. 1993. "How Will We Work in the Year 2000?" *Fortune*, (May 17): 38-52.
Kuttner, R. 1991. *The End of Laissez Faire: National Purpose and the Global Economy after the Cold War*. New York: Alfred Knopf.
Lively, K. 1994. "Big Tuition Increases and Deep Budget Cuts Cause Enrollment to Plummet at California Public Colleges." *Chronicle of Higher Education*, 19: A22.
Lively, K. 1993. "Crawling Toward Recovery: State Support for Public Colleges Up 2% This Year." *Chronicle of Higher Education*, 10: A29.
Luthans, F. and D.L. Bohl. (Eds.). 1993. *Organizational Dynamics*, Special Issue on the Learning Organization, 2: 2-3.
Lynch, D.M., R. Mcferron, L.H. Bowker and I. Beckford 1993. "A Discipline in Trouble: Why More Sociology Departments May be Closing Shortly." *Footnotes*, (February) 21: 3,7.
Mercer, J. and K. Lively. 1994. "The Worst May be Over: Higher Education and the States." *Chronicle of Higher Education*, 18: A25.
Mills, C.W. 1959. *The Sociological Imagination*. New York: Grove Press.
Morgan, G. 1993. *Imaginization: The Art of Creative Management*. Newbury Park, CA: Sage.
Morgan G. 1984. *Images of Organization*. Newbury Park, CA: Sage.
Park, P. 1992. "The Discovery of Participatory Research as a New Scientific Paradigm: Personal and Intellectual Accounts." *The American Sociologist*, 23, 4: 29-42.
Polanyi, K. 1957. *The Great Transformation: The Political and Economic Origins of Our Time*. Boston: Beacon Press.
Reich, R. 1992. *The Work of Nations: Preparing Ourselves for 21st Century Capitalism*. New York: Vintage.
Reich, R. 1983. *The Next American Frontier*. New York: Penguin.

Rossi, P. 1986. "How Applied Sociology Can Save Basic Sociology." *Journal of Applied Sociology*, 3: 1-5.

Rossi, P. and W.F. Whyte. 1983. "The Applied Side of Sociology." in *Applied Sociology*, H.E. Freeman, R.R. Dynes and W.F. Whyte (eds.). San Francisco: Jossey-Bass.

Ruggerio, J.A., and L.C. Weston. 1991. "Working Definitions of Sociological Practice: Results of Recent Practitioners Survey." *Sociological Practice Review*, 1: 64-67.

Sabel, C. and M. Piore. 1984. *The Second Industrial Divide: Possibilities for Prosperity*. New York: Basic.

Senge, P.M. 1990. *The Fifth Dimension: The Art and Practice of The Learning Organization*. New York: Doubleday.

Sharp, L. and J.C. Weidman. 1989. "Early Careers of Undergraduate Humanities Majors." *Journal of Higher Education*, 5: 544-564.

Simpson, R.D. and S.H. Frost. 1993. *Inside College: Undergraduate Education for the Future*. New York: Insight.

Sociology Task Force. 1991. *Liberal Learning and the Sociology Major: A Report to the Profession*. Washington, DC: American Sociological Association.

Toffler, A. 1980. *The Third Wave*. New York: Morrow.

U.S. Department of Labor, Bureau of Labor Statistics. 1993. "The American Workforce: 1992-2005." *Monthly Labor Review*, 11.

Useem, M. 1989. *Liberal Education and the Corporation: The Hiring and Advancement of College Graduates*. New York: Aldine de Gruyter.

Whyte, William F. 1991. "The Social Sciences in the University." *The American Behavioral Scientist*, 5: 618-633.

Womak J., D. Jones and D. Roos. 1990. *The Machine that Changed the World*. New York: Rawson.

Zuboff, Shoshana. 1988. *In the Age of the Smart Machine: The Future of Power and Work*. New York: Basic.

AMERICAN SOCIOLOGICAL ASSOCIATION
DEPARTMENT OF RESEARCH AND DEVELOPMENT

IDEALISTS VS. CAREERISTS:
Graduate School Choices of Sociology Majors[1]

Roberta Spalter-Roth and Nicole Van Vooren

May 2009

INTRODUCTION

This research brief suggests that sociology can attract both idealists and careerists, and that both types of students find a place in graduate school. Sociological skills and concepts are beneficial both to students who go on to graduate school in applied and professional fields and to those who continue on in sociology.

BACKGROUND: THE GROWTH OF OCCUPATIONAL AND PROFESSIONAL DEGREES

The decline of arts and sciences as core disciplines and the corresponding growth of professional programs are important developments in higher education over the last 30 years, according to sociologist Steven Brint and his colleagues (2002; 2005). Occupational or professional degree programs are viewed as providing students with knowledge and skills acquisition that prepares them for the labor market (Council of Graduate Schools 2006; Glazer-Raymo 2004; National Academy of Science 2008). Currently about 60 percent of bachelor's degrees awarded are in occupational/professional fields, including allied health professions, criminology, information systems, human resources, business, and recreation—fields often housed outside schools of arts and sciences (Brint et al. 2005). During this 30-year period, the master's degree, and especially career-oriented, applied, and professional master's degrees became the fastest growing degree among all offered by universities (American Sociological Association, Task Force on the Master's Degree 2009; Council of Graduate Schools, 2006; Strauss 2006). As of 2006, about 85 percent of all graduate degrees awarded were master's degrees.

IMPLICATIONS FOR SOCIOLOGY PROGRAMS

One result of the shift to the professional programs is that sociology undergraduate programs are sending the majority of majors

---

[1] This study has been funded by the Sociology Program of the National Science Foundation. The views stated here reflect those of the authors.

either directly into the workforce or to master's programs in applied sociology or in neighboring or break-away disciplines directly linked to specific careers. We find that the students who enroll in graduate programs, rather than going directly into the labor market, tend to be "careerists" who majored in sociology because it would prepare them for graduate or professional school. Alternatively, those students who enroll in sociology master's programs in sociology tend to be "idealists." However, even those students enrolled in sociology graduate programs are more likely to be in master's programs rather than in PhD programs. The small percentages of those who intend to obtain a PhD also agree strongly that they majored in sociology to prepare them for a graduate or professional degree.

THE STUDY

In spring 2005 the American Sociological Association's Research and Development Department sent an on-line questionnaire to a sample of 1,777 seniors to survey their experiences and satisfaction with the sociology major as well as their future plans for work, graduate school, or both. Early in 2007 we re-surveyed the class of 2005 to learn what they had been doing since graduation. The response rate for the second wave of the survey was 44 percent or 778 graduates.

SECOND-WAVE DATA

In the second wave of the survey we asked about labor market activities and graduate school studies. As part of the labor market information, we asked about job searches, job descriptions, job satisfaction, and closeness of employment to sociological training. Those former sociology majors who were either in graduate school or had completed a post-graduate degree were asked about their discipline and degree level. Thus, we can determine whether or not they entered a sociology program or a more career-oriented program.

In addition, we asked everybody about the skills and concepts they used on the job or in graduate school. Skills included the following: forming a causal hypothesis; using computer resources; using statistical software; evaluating research methods; developing evidence-based arguments; using tests of significance; interpreting the results of data gathering; identifying ethical issues in research; writing reports; and working in diverse groups with others. Concepts included: current sociological explanations about a variety of social issues; social institutions and their impact on individuals; basic theoretical perspectives or paradigms in sociology; basic concepts in sociology (including culture, socialization, institutions, or stratification); important differences in the life experiences of people as they vary by race, class, gender, age, disability and other ascribed statuses; and views of society from alternative or critical perspectives.

PRIOR DATA

In the first wave of the study we had asked about reasons for majoring in sociology. Most students major in sociology because they enjoyed their first course in the subject. The second most frequent reason for majoring is labeled as idealist—students who major because they thought sociology would prepare them to change society, to understand social forces or their own place in society. The third reason for majoring is careerist—students who thought that sociology would prepare them for the job that they wanted or for a professional or graduate degree.

We used SPSS scaling procedures to determine whether the answers to a series of questions about out-of-classroom activities, asked in the first wave, would cluster together into distinctive indexes, each representing a type of social or cultural capital. We found that the activities formed three distinct clusters. These were: (1) Scholarly socialization including membership in a sociology club, participating in the sociological honorary society Alpha Kappa Delta, and attending state, regional, or national sociological meetings; (2) Mentoring activities including student participation in mentoring programs and in faculty research and (3) On-the-job training and job networks including leadership training, participating in internships, engaging in community activities, taking part in service learning programs, and attending job fairs. (The distribution of sociology majors participating in these activities can be seen in *Decreasing the Leak from the Sociology Pipeline: Social and Cultural Capital to Enhance the Post-Baccalaureate Sociology Career* at http://www.asanet.org/galleries/default-file/ASASocPipeBrief.pdf .)

MATCHING THE FIRST AND SECOND WAVE

We were able to match the answers from the first wave of the survey, including

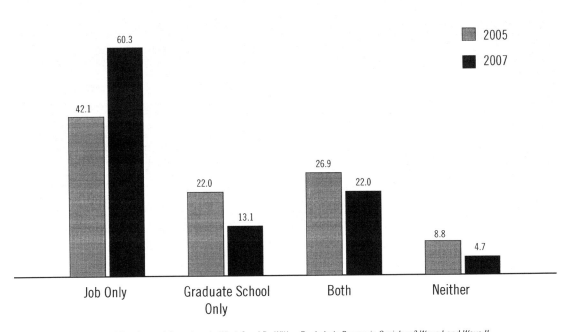

Figure 1. More Sociology Bachelor's Recepients are in the Labor Market
Plans for the Future in 2005 versus Status in 2007

Source: ASA Research and Development Department, *What Can I Do With a Bachelor's Degree in Sociology? Wave I and Wave II*

demographic characteristics, skills and concepts learned, reasons for majoring in sociology, and participation in a range of out-of-classroom activities with the work and graduate school experiences described in the second wave. As a result of this matching, we are able to determine what factors significantly increase the likelihood that sociology majors will enter graduate school, generally, and what factors increase the likelihood that sociology majors will enter graduate school in sociology programs, in particular.

FINDINGS

WHO GOES TO GRADUATE SCHOOL?

Before graduation, we had asked the class of 2005 about their future plans (Spalter-Roth and Erskine 2006). The largest group (42.1 percent) said they intended to find a job. Figure 1 shows that, in fact, by 2007, 60.3 percent were working, 20 percentage points more than initially projected. Although 22 percent planned to go to graduate school and not seek employment in 2005, only 13.1 percent did so in 2007. Finally, 26.9 percent planned to both attend graduate school and seek employment. By 2007, about 22 percent were both in the labor market and in graduate school.

How do the demographic characteristics of those who enrolled in graduate school compare to those who did not? Table 1 compares the characteristics of those former majors who go on to graduate school with those who do not. The table also compares type of undergraduate institution attended, parents' level of education, participation in extra-classroom activities, skills and concepts learned, type of combination majors with sociology, reasons for majoring in sociology, and Grade Point Average (GPA).

We find that the demographic characteristics of sociology majors who went on to graduate school are not significantly different than those who did not. In other words, relatively similar percentages of women, African Americans, Hispanics, and former majors whose parents have less than a college degree enrolled in graduate school and joined the labor force. Although the differences were not significant, a somewhat greater percentage of African Americans and Hispanics were enrolled in graduate school as compared with their percentage in the population of former majors as a whole. Parents' education and the type of institution of higher education that former majors attended were not significantly different for those who enrolled in graduate school and those who did not.

Type of combined major was significant, however, for those who combined psychology and sociology, but not for those who combined sociology and criminal justice. The former were significantly more likely to attend graduate school than were those with other combined degrees or with stand-alone sociology majors, while the latter were not. In other words, joint psychology and sociology majors were significantly more likely to go to graduate school, in contrast with joint criminology and sociology majors, who were more likely to enter the labor force upon graduation.

The GPA of those who enrolled in graduate school was significantly higher than those who did not attend. By contrast, there is no relationship between mastery of skills and

Table 1: Characteristics of Sociology Undergraduate Majors Who Go to Graduate School Compared to those Who Do Not

Gender	Ø
Race	Ø
Type of Undergraduate Institution	Ø
Mother's level of education	Ø
Father's level of education	Ø
Combination Majors	
Criminal Justice	Ø
Psychology	★
Sociology GPA	★
Skills and Concepts	
Research	Ø
Communication	Ø
Conceptual	Ø
Extra Classroom Activities	
On the Job Training & Networking	Ø
Mentoring	★
Scholarly Socialization	★
Reasons for Majoring	
Idealist	★
Careerist	★

★ Statistically significant difference Chi-square or T-test ($p < 0.05$)

Ø Not statistically significant

Source: ASA Research and Development Department, *What Can I Do With a Bachelor's Degree in Sociology? Wave I and Wave II*

the likelihood of continuing one's education beyond the bachelor's level. However, Table 1 does show that those students who participate in some extra-classroom activities, such as mentoring or scholarly socialization activities, are significantly more likely to go on to graduate school. Finally, idealists and careerists are more likely to enter graduate school than those who did not enroll, and these differences are statistically significant.

What happens when we compare these characteristics relative to one another? We used bi-variate logistic regression analysis to answer this question. The independent variables in the model are the following: race and ethnicity, overall GPA, GPA in sociology courses, type of major (joint or single), research skills learned, participation in

> "…those students who participate in some extra-classroom activities, such as mentoring or scholarly socialization activities, are significantly more likely to go on to graduate school…"

mentoring, scholarly socialization, or on-the-job-training and networking activities, and whether or not they were careerists or idealists. The dependent variable was whether or not the former sociology major enrolled in graduate school (see Figure 2).

Sociology majors who enroll in graduate school have much higher GPAs in sociology than their peers who do not enroll in graduate school. When GPA in sociology is included in

Figure 2. Factors Predicting Which Sociology Undergraduate Majors Attend Graduate School, 2007
Odds Ratio

Factor	Odds Ratio
Sociology GPA	8.082
Black/Hispanic	1.787
Careerist Major	1.345

Source: ASA Research and Development Department, *What Can I Do With a Bachelor's Degree in Sociology? Wave I and Wave II*

the model, overall GPA loses significance and drops out of the model. However, Figure 2 shows that a higher sociology GPA increased the likelihood of going to graduate school more than eightfold. The skills and concepts learned as sociology majors are not significant predictors of graduate school enrollment. This lack of significance of skills or concepts learned is probably because those who learned these skills received higher GPA's than those who did not.

Extra-classroom activities available to majors, including on-the-job training and networking activities; mentoring; and sociology socialization, were not significant in predicting who goes to graduate school, despite their significance in the descriptive analysis.

The relative small numbers of Blacks and Hispanics who answered the second wave of the survey were almost twice as likely to enroll in graduate school as are other racial and ethnic groups (namely, whites, Native Americans, and Asians). And, finally, those who go on to enroll in graduate school are significantly more likely to be careerists than those who do not enroll. They major because they think sociology will prepare them for graduate or professional school or perhaps a career. Although majoring for idealistic reasons was significant in the descriptive analysis, it was not significant in the regression model.

In short, GPA in sociology, identifying as African-American or Hispanic, and majoring in sociology for careerist motives significantly increase the odds of sociology majors enrolling in graduate school.

Figure 3. Skills Listed on Graduate School Applications, 2007
Percentage of sociology undergraduate majors reporting listing skills

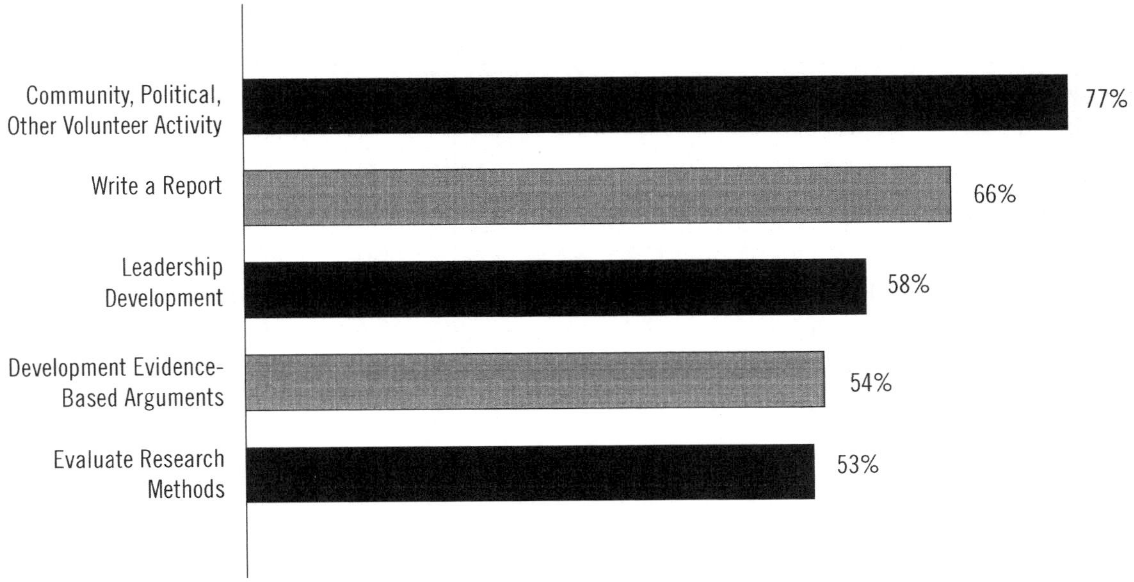

Skill	Percentage
Community, Political, Other Volunteer Activity	77%
Write a Report	66%
Leadership Development	58%
Development Evidence-Based Arguments	54%
Evaluate Research Methods	53%

Source: ASA Research and Development Department, *What Can I Do With a Bachelor's Degree in Sociology? Wave II*

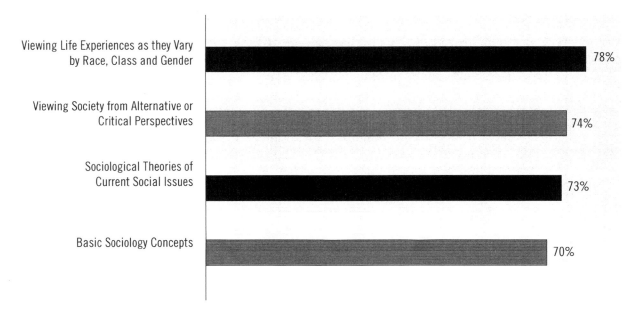

Figure 4. Concepts Used in Graduate Programs, 2007
Percentage of sociology undergraduate majors reporting concepts as being "very useful" in graduate school

Source: ASA Research and Development Department, *What Can I Do With a Bachelor's Degree in Sociology? Wave II*

USING SKILLS AND CONCEPTS IN GRADUATE SCHOOL

The second wave of the survey asked former sociology majors about the skills listed on their applications to graduate school and their ongoing use of skills learned during their graduate programs. Figure 3 shows the skills that at least 50 percent of majors mentioned when applying to graduate school. Community and political activities were the most frequently listed on graduate school applications, perhaps because the majority went into applied fields oriented toward working with clients in particular communities (See *What are They Doing with a Bachelor's Degree in Sociology?* at http://www.asanet.org/galleries/Research/ASAResearchBrief_revised.pdf). Second was the ability to write reports. Third was leadership development. Finally, about half of all majors mentioned research and statistical skills, including developing evidence-based arguments and evaluating the appropriate research method for embarking on a study.

Although survey respondents mentioned the *skills* that they learned as majors in applying for graduate school, it is the *concepts* learned that they reported using on a daily basis in graduate school. In contrast to Figure 3, Figure 4 shows that at least 70 percent of those enrolled in graduate school said that sociological theories and concepts were "very useful" in their graduate coursework. These included understanding the differences in the life experiences of people as they vary by race, class, gender, age, disability, and other ascribed statuses; viewing society from alternative or critical perspectives; knowledge of sociological explanations about current social issues such as crime, racism, poverty, family formation, or religion; and understanding basic concepts in sociology including culture, socialization, institutions, and stratification. The results suggest that these concepts can be used in a wide variety of degree fields in which sociology majors enroll.

CHOICE OF DEGREE FIELDS

About three-quarters of those in graduate school were pursuing master's degrees, about 13 percent were pursuing professional degrees, and about 11 percent were pursuing degrees that would lead to a PhD. Table 2 shows the field of graduate study in which 2005 sociology majors were enrolled in 2007. The largest group was pursuing degrees in sociology (22.4 percent), yet the majority was in other fields (77.6 percent). These fields included social work, education, law, psychology/counseling, other social sciences, business, public policy, and engineering. About 70 percent of undergraduate sociology majors were pursuing degrees in what can be characterized as applied, vocational, or professional-oriented fields.

CHARACTERISTICS OF THOSE WHO CHOOSE SOCIOLOGY IN GRADUATE SCHOOL

Are there differences between those sociology majors who enroll in graduate sociology programs and the much larger number who enroll in the other programs? Table 3 shows that few differences achieve statistical significance. There are no significant differences by type of undergraduate

Table 2: Graduate Fields of Study of 2005 Sociology Graduates, 2007
(in percents)

FIELD OF STUDY	
Sociology	22.4%
Social Work	16.9%
Education	11.4%
Law	10.5%
Other Social Sciences	9.4%
Psychology/Counseling	8.6%
Engineering	7.3%
Business/Management	4.5%
Public Policy/Affairs	3.8%
Other	5.1%
TOTAL	**100%**

Source: ASA Research and Development Department, *What Can I Do With a Bachelor's Degree in Sociology? Wave I and Wave II*

institution, gender, race or ethnicity, or parents' level of education. These two groups are also not distinguished by GPA.

The first set of significant differences (or the lack of them) relates to participation in extra-classroom activities. Majors who go on to enroll in graduate sociology programs are significantly more likely to participate in activities that socialize them into the sociological field. They are more likely than their peers to be selected for sociology honors programs, to participate in Alpha Kappa Delta, and to attend state, regional, or national sociology meetings usually at the instigation of their professors. They are significantly less likely to be joint psychology/sociology majors than regular sociology majors.

The table also shows that those enrolled in sociology programs are significantly more likely to strongly agree that they learned research and statistical skills as undergraduates, including developing evidence-based arguments, evaluating the appropriate research methods for embarking on a study, and mastering statistical computer packages, than those enrolled in other programs. Finally, those enrolled in sociology graduate programs were significantly more likely to have majored in sociology because they were idealists rather than careerists.

In contrast, there are no significant differences in mentoring activities and on-the-job training and networking activities. In other words, sociology faculty members seem equally likely to mentor and include majors who go on to enroll in other graduate fields of study in their research projects as they are those who go on to sociology graduate programs. Sociology majors who enrolled in other graduate programs were as likely to participate in community activities, internships and other on-the-job training and networking activities as those who enrolled in sociology programs. As noted, community activities are the most likely activity to be listed on graduate school applications, and both groups headed to graduate school took advantage of these out-of-the-classroom events.

In what follows, we use bi-variate logistic regression analysis to determine which characteristics are still significant when compared to one another in predicting who goes to graduate school in sociology. As in the previous regression analysis, the independent variables in the model are those that were significant in the cross-tabulations shown in Table 3. These include whether or not they were joint psychology/sociology majors,

> "…those who enroll in sociology graduate programs strongly agree that they have mastered research skills and are more likely to be idealists than careerists. Scholarly socialization, including participation in honors programs, sociology clubs, and sociology meetings…may be helpful in encouraging sociology majors to attend graduate school in their discipline."

whether or not they participated in scholarly socialization activities, whether or not they are idealists, as well as the number of research skills they learned.

Figure 5 shows that all but one of these predictors continues to be significant in the regression model. The number of research skills that respondents strongly agreed that they learned as undergraduates increased the likelihood of going to graduate school in sociology rather than in another program by about 1.3 times. Being an idealist increases the probability of becoming a sociology graduate student by 1.5 times. In contrast, being a joint psychology/sociology major significantly **decreases** the likelihood of attending graduate school in sociology, as these joint majors are more likely to enroll in psychology programs. Scholarly socialization is not significant, however.

In sum, those who enroll in sociology graduate programs strongly agree that they

Table 3: Characteristics of Those Enrolled in Sociology Graduate Programs Compared to Those Enrolled in other Programs, 2007
(in percents)

Gender	Ø
Race	Ø
Type of Undergraduate Institution	Ø
Mother's level of education	Ø
Father's level of education	Ø
Extra Classroom Activities	
On the Job Training & Networking	Ø
Mentoring	Ø
Scholarly Socialization	★
Skills and Concepts	
Research	★
Communication	Ø
Conceptual	Ø
Combination Majors	
Criminal Justice	Ø
Psychology	★
Reasons for Majoring	
Idealist	★
Careerist	Ø
Sociology GPA	Ø

★ Statistically significant difference Chi-square or T-test ($p < 0.05$)
Ø Not statistically significant

Source: ASA Research and Development Department, *What Can I Do With a Bachelor's Degree in Sociology? Wave I and Wave II*

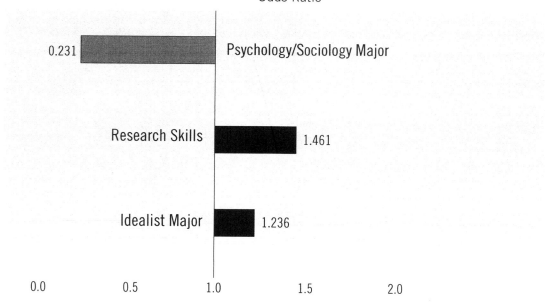

Figure 5. Factors Predicting Which Sociology Undergraduate Majors Pursue Sociology in Graduate School, 2007
Odds Ratio

Source: ASA Research and Development Department, *What Can I Do With a Bachelor's Degree in Sociology? Wave I and Wave II*

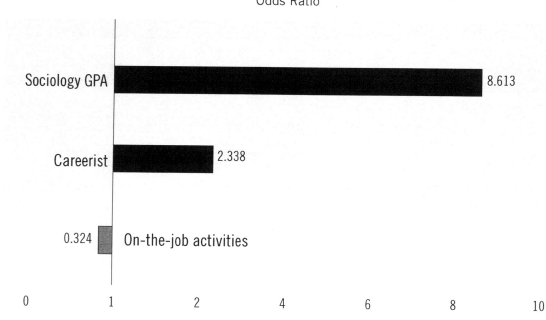

Figure 6. Factors Predicting Which Sociology Undergraduate Majors Enroll in PhD Programs, 2007
Odds Ratio

Source: ASA Research and Development Department, *What Can I Do With a Bachelor's Degree in Sociology? Wave I and Wave II*

have mastered research skills and are more likely to be idealists than careerists. Scholarly socialization, including participation in honors programs, sociology clubs, and sociology meetings, though not significant in the regression analysis, may be helpful in encouraging sociology majors to attend graduate school in their discipline.

ON TO THE PHD

Although a PhD degree is considered the culmination of a sociological education, only 11 percent of this 2005 cohort enrolled in PhD programs. Those in sociology graduate programs are the most likely to be enrolled in a PhD program (29.6 percent), followed by those in psychology programs (22.7 percent), and other social science programs (21.7 percent). Students who enrolled in law school expect a professional degree, but for the remainder, virtually all students are enrolled in master's level programs.

What differences exist between those who enrolled in masters and those who enrolled in PhD programs? There are no significant differences between those who participated in scholarly socialization activities, but there are significant differences in those who participated in mentoring activities and on-the-job training activities. Participating in mentoring and research activities is important for encouraging sociology majors to pursue PhD degrees. On-the-job training activities are significant for those who enroll in master's programs.

When we examine the differences between those who enter a PhD program versus those who do not, we find a negative relationship between participation in on-the-job and networking programs with working towards a PhD degree, in a bi-variate regression analysis (see Figure 6). Mentoring and scholarly socialization are not significant and fall out of the model. Here again, GPA in sociology is significant, while overall GPA drops out of the model. Those with higher undergraduate GPAs are more than eight times as likely to go into a PhD program. Last, PhD students are twice as likely as other respondents to strongly agree that the sociology major prepared them for graduate or professional school.

CONCLUSIONS

At times during its history, sociology has extolled its practical expertise (most recently with an initiative to create a public sociology). This research brief suggests that the sociology major is a gateway to graduate study in professional and applied programs as well as the liberal arts and sciences. Almost 8 out of 10 majors who go on to graduate school do so for degrees other than sociology. Those that enroll in social work, education, counseling, business, and public policy graduate programs believe that the undergraduate sociology major prepares them for their graduate or professional school careers. Half of them listed statistical and methodological skills they had learned on their graduate school applications. At least 70 percent of those who go on to graduate schools find sociology concepts and perspectives, such as social problems, race and gender inequalities, stratification, culture, and critical perspectives, to be very useful in their graduate schooling, regardless of the field. Students who go on to graduate school have higher GPAs in sociology course work, a reflection of their greater mastery of sociological skills, theories, and concepts.

By advertising the sociology major as a coherent framework that results in entry into both practical and academic career pipelines, the student and alumni base should continue to grow. Many students enrolling in applied and professional degree programs already believe this, strongly agreeing that the sociology major is good preparation for graduate and professional school.

Those who do enroll in master's programs in sociology strongly agree that they have mastered research skills including using statistical software packages, interpreting the results of data gathering, and evaluating different research methods. They are likely to be idealists who majored in sociology because they want to change society and to understand social institutions and the relation between social forces and individuals. They have been socialized in the field by belonging to sociology clubs such as Alpha Kappa Delta, and by attending state, regional, or national sociology meetings with their professors. Those enrolled in PhD programs are also likely to strongly agree that the sociology major is helpful for entering graduate or professional school.

In conclusion, sociology attracts majors with an exciting first course, appealing to both idealists and careerists. As noted, both types of students find a place in graduate school. Sociological skills and concepts, learned as undergraduates, are seen as beneficial for graduate school applications and for daily use in graduate school in applied and professional fields and in sociology.

References

American Sociological Association, Task Force on the Master's Degree in Sociology, 2009. *Thinking about the Master's Degree in Sociology: Academic, Applied, Professional, and Everything in Between.* Washington, DC: American Sociological Association.

Brint, Steven. 2002. "The Rise of the 'Practical Arts.'" Pp. 231-59 in *The Future of the City of Intellect: The Changing American University*, edited by S. Brint. Stanford, CA: Stanford University Press.

Brint, Steven, Mark Riddle, Lori Turk-Bicakci and Charles S. Levy. 2005. "From the Liberal to the Practical Arts in American Colleges and Universities: Organizational Analysis and Curricular Change." *The Journal of Higher Education* 76(2):151-80.

Council of Graduate Schools. 2006. *Professional Master's Education: A CGS Guide to Establishing Programs.* Washington, DC: Council of Graduate Schools.

Glazer-Raymo, Judith. 2004. "Trajectories for Professional Master's Education." *Communicator* 37:2(1-2, 5).

National Academy of Science, National Research Council, Committee on Enhancing the Master's Degree in the Natural Sciences. 2008. *Science Professionals: Master's Education for a Competitive World.* Washington, DC: The National Academies Press.

Spalter-Roth, Roberta and William Erskine. 2006. "What Can I Do With a Bachelor's in Sociology? A National Survey of Seniors Majoring in Sociology—First Glances: What Do They Know and Where Are They Going?" *Research Brief.* Washington, DC: American Sociological Association.

Strauss, Valerie. 2006. "More Master's Courses Designed to Expedite Students into Jobs." *Washington Post*, April 18, p. A06.

Research Briefs

The following are links to research briefs and reports produced by the ASA's Department of Research and Development for dissemination in a variety of venues and concerning topics of interest to the discipline and profession. These briefs can be located at
http://www.asanet.org/cs/root/leftnav/research_and_stats/briefs_and_articles/briefs_and_articles
You will need the Adobe Reader to view our PDF versions.

TITLE	YEAR	FORMAT
Sociology Faculty Salaries AY 2008/09: Better Than Other Social Sciences, But Not Above Inflation (**Brief Currently Available to ASA Members Only)	2009	PDF
What's Happening in Your Department: Who's Teaching and How Much?	2009	PDF
Decreasing the Leak from the Sociology Pipeline: Social and Cultural Capital to Enhance the Post-Baccalaureate Sociology Career	2009	PDF
What's Happening in Your Department? A Comparison of Findings From the 2001 and 2007 Department Surveys	2008	PDF
PhD's at Mid-Career: Satisfaction with Work and Family	2008	PDF
Too Many or Too Few PhDs? Employment Opportunities in Academic Sociology	2008	PDF
Pathways to Job Satisfaction: What happened to the Class of 2005	2008	PDF
Sociology Faculty Salaries, AY 2007-08	2008	PDF
How Does Our Membership Grow? Indicators of Change by Gender, Race and Ethnicity by Degree Type, 2001-2007	2008	PDF
What are they Doing With a Bachelor's Degree in Sociology?	2008	PDF
The Health of Sociology: Statistical Fact Sheets, 2007	2007	PDF
Sociology and Other Social Science Salary Increases: Past, Present, and Future	2007	PDF
Race and Ethnicity in the Sociology Pipeline	2007	PDF
Beyond the Ivory Tower: Professionalism, Skills Match, and Job Satisfaction in Sociology [Power Point slide show]	2007	PPT
What Sociologists Know About the Acceptance and Diffusion of Innovation: The Case of Engineering Education	2007	PDF
Resources or Rewards? The Distribution of Work-Family Policies	2006	PDF
Profile of 2005 ASA Membership	2006	PDF
"What Can I Do with a Bachelor's Degree in Sociology?" A National Survey of Seniors Majoring in Sociology—First Glances: What Do They Know and Where Are They Going?	2006	PDF
Race, Ethnicity & American Labor Market	2005	PDF
Race, Ethnicity & Health of Americans	2005	PDF
The Best Time to Have a Baby: Institutional Resources and Family Strategies Among Early Career Sociologists	2004	PDF
Academic Relations: The Use of Supplementary Faculty	2004	PDF
Have Faculty Salaries Peaked? Sociology Wage Growth Flat in Constant Dollars	2004	PDF
Are Sociology Departments Downsizing?	2004	PDF
Sociology Salary Trends	2002	PDF
How Does Your Department Compare? A Peer Analysis from the AY 2000-2001 Survey of Baccalaureate and Graduate Programs in Sociology	2003	PDF
Graduate Department Vitality: Changes Continue in the Right Direction	2001	PDF
Minorities at Three Stages in the Sociology Pipeline	2001	PDF
The Pipeline for Faculty of Color in Sociology	2001	PDF
Profile of the 2001 ASA Membership	2001	PDF
Use of Adjunct and Part-time Faculty in Sociology	2001	PDF
Gender in the Early Stages of the Sociological Career	2000	PDF
New Doctorates in Sociology: Professions Inside and Outside the Academy	2000	PDF
After the Fall: The Growth Rate of Sociology BAs Outstrips Other Disciplines Indicating an Improved Market for Sociologists	1998	PDF
Update 1: After the Fall: Growth Trends Continue		PDF
Update 2: BA Growth Trend: Sociology Overtakes Economics		PDF

The Future Lives of Sociology Graduates

NEIL GUPPY, KERRY GREER, AND NICOLE MALETTE
University of British Columbia

KRISTYN FRANK
Statistics Canada

After earning a bachelor's degree, the fate of sociology undergraduates is mysterious. We know little about how many Canadians have sociology degrees and even less about what they do after graduation. Using a variety of data sources from Statistics Canada, we paint a historical portrait of the discipline by charting how many people, and who, become sociologists. We also examine where they are employed, how much they earn, and how satisfied they are with their degrees. Our findings show that sociology graduates are competitive with other fields of study on both wages and employment, but are less satisfied with their degrees. We stress how the discipline would be wise to pay more attention to our graduates and their opportunities.

Nous en savons peu sur l'avenir et le nombre de Canadiens ayant des diplômes de sociologie et encore moins ce qu'ils font après la graduation. Grâce à des données de Statistiques Canada, nous dressons un portrait historique de la discipline en montrant combien de personnes, et quel genre de personnes, deviennent sociologues. Nous examinons aussi où ils travaillent, combien ils gagnent et leur niveau de satisfaction par rapport à leurs diplômes. Nos résultats démontrent que les gradués en sociologie sont compétitifs par rapport aux autres champs d'étude en termes de salaire et de taux d'emploi, mais qu'ils sont moins satisfaits de leurs diplômes. Nous notons que la discipline aurait intérêt à porter plus d'attention sur les gradués et leurs opportunités.

The research reported here was financially supported by the Canadian Sociological Association. We are also grateful to Statistics Canada for providing us with the data used in this paper. We thank Amy Swiffen for launching this project. We received excellent comments from Howard Ramos, Simon Langlois, Edward Grabb, and Jim Conley on an earlier version of this paper.

Neil Guppy, Department of Sociology, University of British Columbia, Vancouver Campus, 6303 NW Marine Drive, Vancouver, BC, Canada V6T 1Z1. E-mail: neil.guppy@ubc.ca

© 2017 Canadian Sociological Association/La Société canadienne de sociologie

THE AFTERLIFE OF A SOCIOLOGY BA graduate remains mysterious. Upon receipt of a sociology credential, our alumni virtually disappear from our radar. Media releases provide occasional blips. Many know that Stephan Dion, the former leader of the Liberal Party and a current federal Cabinet Minister, has a PhD in sociology, and likewise that Monique Bégin, a former prominent Cabinet Minister, also has degrees in sociology. Many will also know that the pollster Angus Reid has a PhD in sociology. Others will have read that the sociology scholars Danielle Juteau and Peter Li, for example, were appointed to the Order of Canada in 2013 and 2016, respectively.

Few sociology degree holders attain this high public profile. No television shows, blockbuster movies, best-selling books, or trending videos feature sociologists. For the most part, mystery surrounds the life trajectories of Canadians with degrees in sociology. This mystery, however, does not deter social commentary about sociology degrees and their recipients. Margaret Wente (2012a, 2012b), columnist for *The Globe and Mail*, railed that sociology students had been "sold a bill of goods" with degrees that are "increasingly worthless."

The invisibility of sociology degree holders makes countering, or confirming, Wente-like assertions difficult. No reliable source specifies the number of Canadians who are trained to "commit sociology," although estimations are not difficult. We know from Statistics Canada reports, and associated projections, that there are currently around 200,000 Canadians who have graduated from Canadian universities with sociology BAs Mortality will have trimmed the number of living graduates slightly, but if we include both immigrants trained abroad and Canadians who earned undergraduate sociology degrees in foreign schools, this approximation would seem robust. The number might also be marginally higher if we included students who did not do their first degree in sociology, but subsequently earned a master's or PhD in the discipline. Finally, of course, many people who received a first degree in sociology subsequently completed degrees in other fields, including law, planning, and teaching. Although they are counted in the 200,000 sociology BA graduates noted above, they are frequently lost to the discipline, and social commentators, because they self-identify as lawyers, planners, or teachers.

A different perspective on the preponderance of sociologists comes from examining the growth among degree recipients. In 1974, there were 74,851 undergraduate degrees awarded by Canadian universities, a figure that grew to 124,861 in 1998 (Education in Canada [Statistics Canada various years]). This represents a growth of 40.0 percent. In contrast, sociology degree recipients rose by 58.4 percent.

The picture of disciplinary growth is bleaker recently. In 1974, the 2,350 sociology degrees granted represented 3.1 percent of all undergraduate degrees awarded, a percentage that grew to 4.5 by 1998. Subsequently, the proportion of sociology degrees has dropped precipitously to 1.7. Part of this decline is possibly artifactual, a consequence of our having to use two different data sources. However, if we use a single data source and contrast data from 2000 and 2013 the drop in the percentage of sociology degrees as a fraction of all degrees declines from 3.0 to 1.7 percent. Explaining this recent drop likely involves a combination of several factors, including perhaps a slowing of growth in public sector employment where many sociologists have historically found work, certainly the competing mix of disciplines now on offer in Canadian universities, and maybe the continuing bad press of sociology as a discipline.

Our goal is not, however, simply to estimate and explain trends in the number of Canadians with sociology degrees. We seek as well to provide a profile of who they are, what they do, how they fare in the workforce, and what they self-report about the value of their degrees. Our paper updates earlier work on graduates of the discipline (e.g., Davies, Mosher, and O'Grady 1992; Davies and Walters 2008).

METHODS

We rely on official statistics from five principal Statistics Canada sources—various years of the Canadian census, the 2011 National Household Survey (NHS), the National Graduates Survey (NGS), annual profiles of higher education (e.g., Education in Canada), and special tabulations provided by the Education Directorate of Statistics Canada. Based on the 1991 and 2001 Census, and the 2011 NHS, we are able to provide activity patterns for people who report their highest degree as being sociology. The long form Census and NHS questionnaires asked the following: "What was the major field of study of the *highest* degree, certificate or diploma that this person completed?" (emphasis in original). This question about major field of study (MFS) for a person's highest postsecondary degree excludes, by definition, anyone trained in sociology who goes on to complete another degree, for example in law, planning, or teaching, and who chose to report that latter degree as their "MFS." We think that the likelihood of the latter is high. Especially, for more recent cohorts, where more and more people have pursued multiple university degrees, the census figures will underestimate the *absolute* number of sociology degree holders, although not necessarily muddying the *proportionate* estimates of who these people are and what they might now be doing.

The NGS is helpful because we can identify individuals whose MFS was sociology, even if they have completed a further degree, which they might define as their "MFS." The 2013 NGS asked respondents who graduated in 2009 to 2010 about all the postsecondary degrees they had attained. The NGS has a very small sample size, which means we can identify only a very few sociology degree recipients. Respondents self-reported on their satisfaction with their degrees, their earnings, and their occupations.

Exactly how field of study is defined has ramifications for our results. In the census and the NHS, Statistics Canada has used two field of study classifications during the 1991 to 2011 interval. For the 1991 and 2001 Censuses, the "MFS" classification was used and the 2011 NHS uses a field of study categorization known as CIP, or Classification of Instructional Programs. Both classifications categorize education courses and learning experiences into discrete clusters that map onto occupational qualifications, preparation for advanced degrees, or programs of study in defined areas (Statistics Canada 2011). Sociology, coded 451101 in the CIP, falls under the social sciences umbrella and is distinct from rural sociology (451401), sociology and anthropology (451301), urban studies (451201), and demography and population studies (450401). Although the coding scheme changed in 2011, an MFS-CIP concordance table helps to ensure that the classification of individuals to the "sociology" field of study is congruent over time. In addition, this study uses the 1991 and 2001 Harmonized Census Microdata files that were developed with the primary objective of creating data with historically comparable census variables.

DEMOGRAPHIC COMPOSITION OVER TIME

Of the approximately 200,000 Canadians who have degrees in sociology, a strong majority are women. The upper panel of Table 1 shows the sex composition of Canadians reporting in the census (or in 2011, the NHS)

Table 1

Percentage of Women, Francophone, and Visible Minority Sociology Degree Holders at Time of Survey by Age Group (1991, 2001, and 2011)

	65–74	55–64	45–54	35–44	25–34	Total
Women						
1991	74.4	67.3	57.1	64.8	72.4	67.2
2001	60.0	59.0	66.3	73.0	74.0	70.3
2011	63.1	65.5	70.5	73.6	75.8	71.7
Francophone						
1991	2.4	3.4	7.6	9.5	8.3	8.4
2001	4.6	6.8	8.7	8.9	4.6	6.9
2011	8.2	9.8	9.0	6.6	5.7	7.4
Visible minority						
1991	6.5	8.5	11.0	8.4	13.9	11.2
2001	12.8	13.0	10.5	16.8	16.4	14.6
2011	13.8	12.3	19.0	19.1	21.1	18.1

Sources: 1991 and 2001 Harmonized Census Microdata and 2011 National Household Survey, Statistics Canada.

that they hold a sociology degree as their main credential. As the rightmost column clearly shows, slightly more than two-thirds of people reporting a sociology credential are women, a percentage that has increased slightly across the three time periods. The same basic pattern holds when we compare across age groups, with the older cohorts typically composed of a slightly smaller proportion of women (seen by comparing down the columns). The pattern is not perfectly linear as the very oldest cohort, those aged 65 to 74 in 1991 and born just after the First World War, represent the cohort with the second highest proportion of female sociology degree holders.

The robustness of the basic pattern of feminization shown in Table 1 is corroborated by examining the annual share of women who graduated from Canadian universities with sociology BA degrees between the early 1970s (when 58 percent of degree recipients were women) and more recently in the 2010s (75 to 80 percent women; Education in Canada and special tabulations purchased from Statistics Canada). The percentage of women among sociology graduates is now between 75 and 80 percent at the undergraduate level, while it is about 70 percent at the MA level, and at about 60 percent among PhD students.

Several explanations for the feminization of sociology as a field of study seem plausible (Roos 1997). Part of the increase in the absolute number of women trained as sociologists is a function of the increasing percentage of women graduating from university generally which between 1976 and 2008 grew from 48 to 61 percent (Davies and Guppy 2014:152). This helps to explain a rise in the absolute number of women, but it does not account for the proportionate increase among women pursuing sociology. One explanation for rising feminization among sociology graduates is that women select sociology as a field of study because of its personal resonance. The discipline continues to place more stress than others do on issues about which women care deeply, including social justice, children and family, sexuality, and gender equity (for an economics/sociology comparison, see Fourcade, Ollion, and Algan 2015). Alternatively, social science degrees, at least outside of economics, have never led to particularly high-paying jobs, jobs to which men have traditionally been drawn and from which women have been restricted (see the comparative income data by field and sex that we report next).

The middle panel of Table 1 shows trends among Francophones and Anglophones. A first impression is that the number of Francophones reporting sociology as a main undergraduate degree is lower than expected. Using as a baseline the approximate percentage of Canadians who report French as their home language (the measure of language used in the table), the Francophone proportion has been in the 20 to 22 percent range for the past few decades. By comparison, the figures in Table 1 show that among people aged 25 to 34 in 2011, only 5.7 percent of sociology degree holders are Francophones. This finding was unexpected as we initially interpreted the Dion and Bégin examples with which we began the paper as emblematic of a strong pool of sociology graduates among Canadian Francophones.

Historically, it appears that there was an uptick in the percentage of Francophones holding sociology degrees among the older cohorts, but this pattern has reversed more recently. For the three oldest cohorts in Table 1, greater percentages of Francophones reported holding sociology degrees in each successive decade from 1991 to 2011. That trend is sharply different compared to the pattern among 25- to 34-year-olds. In 2011, a lower percentage of young people (5.7 percent) were reported holding their highest degrees in sociology as compared to the same-age cohort in 1991 (8.3 percent).

Another way to make these comparisons is to examine the percentage of Francophones and Anglophones that hold university degrees in fields other than sociology (not reported in a table). Among the youngest cohort in 2011 (aged 25 to 34), 17.1 percent of all nonsociology degree holders were Francophones, but only 5.7 percent of sociology degree holders were Francophone.

Where this pattern is sharply different is among those who hold graduate degrees in sociology. When we examine the proportion of people with master's or doctoral degrees, the percentage of people reporting their home language as French rises to 25.3 in the 2011 cohort (data not presented in a table). Francophone speakers appear less likely to hold an undergraduate degree than one might expect, but a higher percentage than expected hold graduate degrees (like the degrees possessed by Bégin and Dion).

Part of the story here might be that more Francophones completing a BA in sociology proceed to other degrees (e.g., in law, planning, or teaching), than is the case among Anglophones. The theme of personal resonance, which we suggested explained some of the rise in feminization, might also be at work here in parsing out the Anglophone-Francophone contrast. That is, Francophone sociology has traditionally focused on issues of identity and "distinct society" (Rocher 1970; Warren 2009), issues that were more important among the older cohorts where we saw increases among Francophones with sociology degrees. However, these issues have apparently not galvanized younger Francophones in more recent years, with the result that sociology may have lost some of its appeal as a field of study.

There is also a very different explanation that is more structural in nature. Rather than explaining these patterns with respect to personal choice, it could be that the mix of disciplines in Québec differs significantly from the rest of Canada. For example, programs of study examining regionalism and nationalism as separate disciplines have long been more common in Québec than elsewhere. International studies, as well as business and administration studies, have begun to eclipse Québec studies in the curriculum. These "competing" disciplines may be attracting Francophone students who might otherwise have been attracted to sociology.

A third demographic pattern we explore is the relative number of sociology credentials reported by visible and nonvisible minority Canadians. As context, recall that in recent years Statistics Canada shows that members of visible minorities have reported higher percentages of university degrees (43 percent among 24- to 44-year-olds in 2011) than have nonvisible minority Canadians (28 percent), a pattern that is more pronounced for men than for women (the gap is 19 percent for men and 12 percent for women). We also know that fields of study are chosen by significantly different percentages of visible and nonvisible minorities, with visible minority students more frequently choosing STEM (science, technology, engineering, and mathematics)-related fields (Dickson 2010; Zarifa 2012).

The bottom panel of Table 1 shows a relatively clear pattern with a growing number of visible minority Canadians earning sociology degrees. This is evident first from the rightmost column, comparing figures for 1991 (11.2 percent), 2001 (14.6 percent), and 2011 (18.1 percent). A second way to see this pattern is to note that among older age groups, the percentage of sociology degree holders is always lower than it is among the younger cohorts (those from 1991). Remember, however, that this is the proportion of people with sociology degrees who are members of a visible minority. Given that about 40 percent of Canadians who are visible minorities hold a university degree, these lower percentage figures tell us that sociology is not a preferred field of study for members of visible minority populations, even if the percentage is increasing.

Over time, members of visible minorities in Canada have increased substantially their proportions of higher education credentials, but visible minorities have not been as attracted to sociology as to some other disciplines, such as economics and computer science. Some observers suppose that this is more a function of academic preparation. Asian students in particular are more prepared to succeed in STEM fields where they are attracted more frequently (Dickson 2010). Some other U.S. studies attribute the field of study choices of Asian-Americans to earnings differences across fields (e.g., Leong and Tata 1990; Liu 1998; Xie and Goyette 2003).

These explanations focus on the lower percentage of visible minority students completing sociology degrees, but they do not account for the relative growth among visible-minority students who do chose to graduate in sociology (a growth of over 60 percent between 1991 and 2011). As with both gender and language group, it might be personal resonance that is at work here; that is sociology is one of the disciplines most likely to address issues of racialization, migration, and ethnic identity. This focuses on personal choices of disciplines links directly to labor force outcomes and leads us to examine the activity patterns of Canadians with sociology degrees after graduation.

LABOR FORCE AND OTHER ACTIVITY

Some commentators have suggested that sociology graduates would have been wiser to have swapped "sociology for socket sets," ensuring their preparedness for meaningful employment (Fletcher 2012). This implies that the unemployment rate among sociology graduates should be above the national average if the alleged benefit of "swapping sociology for socket sets" holds any truth. However, the unemployment rates in both 1991 and 2011 for sociology degree holders, 6.1 and 5.0 percent, respectively, are below the national average for the entire labor force (8.6 and 6.1 percent for all workers for the respective years, ages 25 to 60; Census/NHS).

A tougher test comes in comparing unemployment rates for all bachelor degree holders (excluding sociology) with sociology degree holders. The rates in 2011 were 4.9 and 4.5 percent, respectively. In short, sociology degree recipients are as likely, if not even slightly more likely, to be employed than are nonsociology bachelor degree holders.

Table 2 presents the employment status of women and men whose highest degree was sociology, for two age cohorts for both 1991 and 2011. For people aged 25 to 60 rates of employment are more than 80 percent, and higher for men than women. A profile of employment status in more recent times comes from examining the narrower age range of 30- to 39-year-olds. Here too, we find that the employment rates are lower and the unemployment rates higher for women, relative to the broader age cohort, a finding we attribute to motherhood experiences for women in this age range. Motherhood responsibilities also presumably account for the greater likelihood of women than men not to be in the labor force (the bottom row of the table).

Table 2

Employment Status of Bachelor's Degree Holders in Sociology by Age and Year

	Aged 25–60				Aged 30–39			
	1991		2011		1991		2011	
	♀ (%)	♂ (%)	♀ (%)	♂ (%)	♀ (%)	♂ (%)	♀ (%)	♂ (%)
Employed	80.8	88.8	81.2	86.3	77.8	89.7	82.9	91.7
Unemployed	5.8	6.7	4.9	5.2	7.2	6.6	4.6	4.3
Not in labor force	13.4	4.6	13.9	8.5	14.9	3.8	12.5	4.0
Total	100.0	100.0	100.0	100.0	100.0	100.0	100.0	100.0

Sources: 1991 Harmonized Census Microdata and 2011 National Household Survey, Statistics Canada.

One main message of Table 2 is that sociology degree recipients do not pay an unemployment penalty. In part, this is a function of the asset that any university credential holds, and is not necessarily a specific benefit of sociology as a chosen field of study. Young bachelor degree holders across all fields of study tend to receive higher earnings and are more likely to be employed full year, full time than are high school graduates (Frank, Frenette, and Morissette 2015). Nevertheless, these data patterns support the view that a sociology degree is superior to choosing a line of employment where a degree is not a requirement (as in auto mechanics or other "socket set" lines of work).

Table 3 reports the top 10 occupations in which sociologists aged 25 to 39 reported working in 2011. For women, the range of jobs includes both caring professions (community service work and social work) and human resource professions (administrative and human resource roles). Many of these jobs are presumably in the public sector, but a range of mainly private sector jobs would also be included (e.g., retail and wholesale managers, retail sales). These jobs are reasonably stable across the data we have for 1991, 2001, and 2011 as well as being fairly consistent for older cohorts as well. For men, professions in the criminal justice system are notable (e.g., police, corrections), although a higher proportion than

Table 3

Top 10 Occupations of Bachelor's Degree Holders in Sociology by Sex, Age 25 to 39 (2011)

Women	Men
Social and community service workers	Police officers (except commissioned)
Administrative assistants	Social and community service workers
Administrative officers	Correctional service officers
Social workers	Retail salespersons
Elementary school and kindergarten teachers	Corporate sales managers
Retail and wholesale trade managers	Sales and account representatives—wholesale trade (nontechnical)
Human resources professionals	Technical sales specialists—wholesale trade
Early childhood educators and assistants	Banking, credit, and other investment managers
Professional occupations in advertising, marketing, and public relations	Employment insurance, immigration, border services, and revenue officers
Retail salespersons	Computer network technicians

Source: 2011 National Household Survey, Statistics Canada.

we anticipated seem to be in the private sector (e.g., retail and corporate sales, finance). As for women, these occupational groups are fairly consistent over time and across older cohorts, and the patterns are similar to reports from the United States (see Senter, Spalter-Roth, and Van Vooren 2015). Although the cell sizes diminish very quickly with different categorizations, the general patterns seem reasonably similar when comparing occupational destinations for visible and nonvisible minorities as well as by home language group. The latter patterns suggest that ethnicity and language group do not appear to alter the typical occupational profiles.

We also use data from the 2011 NHS to examine the median earnings (in 2010 dollars) of sociologists as compared to others with bachelor's degrees in the social sciences. We include only people employed full year, full time with positive earnings and no self-employment income. As Table 4 reveals, the earnings of sociologists are typically close to, although usually below, the median earnings of those with other social science degrees (for findings on social science versus other broad fields of study, see Finnie et al. 2016). Sociologists have lower median earnings on all comparisons with economists and almost all comparisons with political science graduates, but sociologists routinely have earnings that are above peers who graduated with bachelor's degrees in history or psychology. Sociologists report wage and salary earnings broadly in keeping with their colleagues who chose other cognate fields of study. Recall again, however, that these findings relate to people whose highest degree was reported on the census as sociology, and so the earnings reported here exclude people who trained first in sociology but then went on to earn a second degree in a different field (e.g., law, planning, or teaching). This is true too, of course, for graduates in other fields of study reported in Table 5, so our comparison is "apples to apples."

We pursue the earnings issue further in Table 5 by examining wage and salary data for individual sociologists who chose to pursue graduate study. Is it worth it to earn a graduate degree? The answer to this question depends on a person's level of degree and gender. For those with an undergraduate degree, the pattern of lower female earnings is replicated across the three age cohorts in the table. For women who earn an MA degree, the higher credential pays off in the labor market with increased earnings across all age cohorts. For men, however, earning the MA degree is not worth the costs, at least in terms of earnings relative to their colleagues who chose to stop their studies with the BA. For men with an MA, they not only earn less than males with only a BA, but men with an MA routinely earn less than women with an MA. At the PhD level, our comparisons are less robust because cell sizes are smaller (and for young men insufficient to report earnings), but the earnings are higher as would be expected.

Judging the robustness of the previous earnings results requires a more complex specification of the models used to examine annual wages and salaries. In Table 6, we follow conventional practice in examining

Table 4

Median Annual Earnings (Wage and Salary) for BA Degree Holders by Field of Study, Gender, Ethnicity, and Language Group (2011)

Field of study	Gender		Ethnicity		Language	
	Women	Men	Visible minority	Nonvisible minority	Francophone	Anglophone
Economics	55,259	72,885	52,400	75,000	61,062	73,774
History	51,259	61,038	49,408	57,800	52,392	58,076
Political Science	58,172	67,836	52,304	66,061	58,162	65,800
Psychology	52,921	63,266	50,014	56,409	49,563	56,903
Sociology	54,186	68,779	51,613	59,732	56,000	58,968
All Social Science (excluding sociology)	55,548	70,000	52,599	66,749	57,787	66,852

Source: 2011 National Household Survey, Statistics Canada.
Note: Earnings refer to 2010 calendar year.

Table 5

Median Annual Earnings (Wage and Salary) of Sociology Graduates by Highest Degree, Sex, and Age Group (2011)

Highest degree	25–34 years		35–44 years		45–54 years	
	Women	Men	Women	Men	Women	Men
Bachelor's	44,684	51,893	60,176	76,373	65,766	79,590
Master's	49,108	47,118	64,210	63,886	72,328	70,480
PhD	62,996	N/A	67,223	81,150	100,000	100,000

Source: 2011 National Household Survey, Statistics Canada.
Note: Earnings for 2010.

Table 6

Median Regression Results for Annual Earnings (Wages and Salaries) of Full-Time Workers with Sociology Degrees, Aged 25 to 54 (2011)

	Model 1		Model 2	
	Coefficient	SE	Coefficient	SE
Age	4,277.62***	578.10	4,454.75***	560.12
Age squared	−42.71***	7.81	−45.17***	7.58
Weeks worked	1,024.73***	37.22	1,020.43***	32.28
Male (female = ref. cat.)	10,255.45***	1242.78	9,553.93***	1,242.71
Nonvisible minority (visible minority = ref. cat.)	6,760.18***	1057.37	6,438.71***	1,029.20
English/other language (Francophone = ref. cat.)	7,000.81***	1124.51	7,678.78***	1,275.30
Graduate degree (Bachelor's degree = ref. cat.)			3,622.23**	1,208.55
Pseudo-R^2	0.157		0.158	
N		6,344		

Source: 2011 National Household Survey, Statistics Canada.
Notes: ***$p < .001$; **$p < .01$; graduate degrees include both MA and PhD. Earnings for 2010.

annual wages and salaries, controlling for age (to account for earnings increasing with years of experience), age squared (the plateauing of earnings at upper ages), and weeks worked (employment effort). To take into account the skewed earning distributions, and to stay consistent with Table 6, we use quantile (median) regression models.

In Model 1, we simultaneously introduce sex, ethnicity, and home language. The findings show that earnings increase by about $4,300 for every year of age, although the rate of that increase declines with age (see the negative coefficient for age squared). For every extra week worked, people received about $1,000 more. The basic earnings patterns of Table 4 are repeated in this first model. Net of other factors men with a sociology degree earn about $10,200 more than do women, nonvisible minority members earn about $6,800 more than do members of visible minorities, and Anglophones earn about $7,000 more than do Francophones. The latter wage gap may also partially account for the lower attraction to sociology among Francophones (see Table 2). In Model 2, where we introduce the possession (or lack) of a graduate degree, we can determine more precisely whether gaining a postgraduate credential pays off economically. The answer is yes, with a reservation. Net of other factors, a graduate degree is worth about $3,600 more each year. The reservation is that "economic payoff" depends on how you calculate opportunity costs. Over the course of a career, the $3,600 annual increment would not likely cover the foregone earnings for someone who took eight years to complete a PhD (post BA). The cell sizes are not sufficiently large to allow us to adjudicate on the difference between the MA and PhD degrees.

Finally, we turn to self-report data from the 2013 NGS to examine how satisfied recent sociology graduates are with their degree choice, occupation, and employment income. As shown in Table 7, among graduates in the six fields of study considered, sociology graduates are the least likely to say that they are "satisfied" or "very satisfied" with their degree choice (44 percent). In addition, although about two-thirds (68 percent) indicate that they are satisfied or very satisfied with their income, this figure is the lowest of all the fields of study that are compared. On the other hand, sociology graduates are relatively happy with their occupations, ranking near the top in this regard, with 94 percent saying that they are satisfied or very satisfied. The contrasts with graduates from other fields of study

Table 7

Percentage of 2010–2011 Graduates Satisfied with Their Degree Choice, Occupation, and Employment Earnings by Field of Study (2013)

Field of study	Percent satisfied or very satisfied with		
	Degree choice	Occupation	Earnings
Anthropology ($N = 69$)	54	74	74
Economics ($N = 144$)	50	84	74
Political Science ($N = 215$)	57	90	74
Criminology ($N = 75$)	73	96	94
Engineering ($N = 1,487$)	86	93	86
Sociology ($N = 193$)	44	94	68

Source: National Graduates Survey, 2013.

in the social sciences are typically not large although sociologists tend to be among the least satisfied about their field of study.

CONCLUSION AND DISCUSSION

Although the sociology discipline has been the butt of sniping in the popular press regarding the value of the degree, these critiques are generally not supported by the evidence. In particular, the findings reported here regarding labor force participation show that those with sociology degrees clearly hold their own in terms of both employment rates and earnings. Where the critiques have merit, at least to some degree, is in some of the evidence concerning recent self-assessments of sociology graduates regarding their degree. Social commentary in the media may exaggerate the woes of sociology graduates, but this popular discourse is not entirely out of step with some of the self-reports of our most recent graduates. How much this media commentary influences the attitudes of sociology degree holders is open to conjecture, but the self-report data about satisfaction levels is less sanguine about the discipline than the behavioral evidence on employment rates and earnings might lead one to expect.

In light of the rather sharp declines in the last decade among the percentage of students choosing to graduate with degrees in sociology, at a time when university degrees are in increasing demand, this is not a positive story for the discipline. Our data on earnings, occupations, and employment rates all show that sociology graduates do reasonably well in the market place. The more positive message that should emerge from these behavioral indicators may be eroding, however, because some media pundits and some of our own graduates currently see the discipline in comparatively negative terms. Certainly in the last decade, students are voting with their feet by seeking other degrees in which to major.

As for historical sweep, we have shown clearly that the discipline was attractive to ever greater proportions of students up until the turn of the twenty-first century. Especially for women, sociology was a degree of choice. A growing proportion of visible minority group members were also choosing sociology, even if in relative terms sociology was not as attractive to them as the STEM fields. Among Francophone students, sociology has not been as enticing as it has been for Anglophones. We suspect that some of the patterns we have found, especially for Francophones and Anglophones, are attributable to differing structural alignments in higher education, especially in how fields of study are organized in Québec compared to the rest of Canada. The recent rise across Canada of new fields of study, in such areas as women's studies and globalization, for example, has likely attracted some students who would in previous generations have chosen sociology.

We also posited, consistent with the research literature, that beyond structural change some of the ebb and flow of sociology student numbers could be attributable to active choices that students made about issues that resonated with them. For students in sociology at least, the choice of fields in this sense may be less driven by financial matters and career prospects than by the topic for which they felt a greater connection or empathy. Based on this logic, the recent decline in the attractiveness of sociology as a major may thus be due to students either taking on a more utilitarian perspective when it comes to career choice, or to students seeing the discipline as increasingly out of touch with their interests (or both).

Sociology faculty need to do more to highlight for our students that the conceptual and methodological disciplinary skills provided by sociology can benefit graduates in their subsequent careers, inside and outside the labor market. It is important to link the classroom with the contributions that our graduates can make to both civil society and labor market. Students also have an obligation to work to build their careers while at university, through both extracurricular activities and coursework. Instructors can play a facilitative role in helping students to make connections between their university experiences and life beyond the BA.

Sociology departments also have an important role to play. More work needs to be done to highlight the careers that can be pursued, as well as the community contributions that can be made, by individuals who have graduated with a sociology degree. One place this could easily occur is through the Web pages of every department in the country, where the "future lives" of graduates can and should be profiled. There are also opportunities for departments to work more closely with alumni and development offices to maintain contact with graduates. Finally, there is an opportunity for the profession to build on the individual newsletters, twitter feeds, and other social media promotions that many departments produce. These initiatives would help in mounting a more coordinated effort to promote the discipline not only to our sociology alumni but to others as well.

References

Davies, S. and N. Guppy. 2014. *The Schooled Society*. 3d ed. Toronto: Oxford University Press.

Davies, S., C. Mosher and B. O'Grady. 1992. "Canadian Sociology and Anthropology Graduates in the 1980's Labour Market, Part I: 1986 Graduates in 1988." *Society/Societe* 16(1): 1–6.

Davies, S. and D. Walters. 2008. "The Value of a Sociology Degree." Pp. 10–18 in *Society in Question*, 5th ed., edited by R.J. Brym. Toronto: Thomson Nelson.

Dickson, L. 2010. "Race and Gender Differences in College Major Choice." *The Annals of the American Academy of Political and Social Sciences* 627:108–24.

Finnie, R., D. Afshar, E. Bozkurt, M. Miyariri and D. Pavlic. 2016. *Barista or Better? New Evidence on the Earnings of Post-secondary Education Graduates: A Tax Linkage Approach*. Ottawa: University of Ottawa, Education Policy Research Initiative.

Fletcher, T. 2012. "Swapping Sociology for Socket Sets." *Surrey Leader*. Retrieved February 8, 2016 (http://www.surreyleader.com/opinion/171040241.html).

Fourcade, M., E. Ollion and Y. Algan. 2015. "The Superiority of Economics." *Journal of Economic Perspectives* 29(1):89–114.

Frank, K., M. Frenette and R. Morissette. 2015. "Labour Market Outcomes of Young Postsecondary Graduates, 2005 to 2012." Economic Insights, No. 50. Statistics Canada Catalogue no. 11-626-X. Ottawa: Statistics Canada.

Leong, F.T.L. and S.P. Tata. 1990. "Sex and Acculturation Differences in Occupational Values among Chinese-American Children." *Journal of Counselling Psychology* 37(2):208–12.

Liu, R.W. 1998. "Educational and Career Expectations of Chinese-American College Students." *Journal of College Student Development* 39(6):577–88.

Rocher, G. 1970. "L'Avenir de la sociologie au Canada." Pp. 14–29 in *The Future of Sociology in Canada*, edited by J. Loubser. Montreal: Canadian Sociology and Anthropology Association.

Roos, P. 1997. "Occupational Feminization, Occupational Decline? Sociology's Changing Sex Composition." *The American Sociologist* 28(1):75–88.

Senter, M., R. Spalter-Roth and N. Van Vooren. 2015. *Jobs, Careers and Sociological Skills: The Employment Experiences of 2012 Sociology Majors*. Washington, DC: American Sociological Association.

Statistics Canada. 2011. "Classification of Instructional Programs (CIP) Canada." Catalogue No. 12-590-X. Ottawa: Ministry of Industry.

Statistics Canada. various years. "Education in Canada." Catalogue No. 81-229-XIB. Retrieved April 2, 2017 (http://publications.gc.ca/Collection-R/Statcan/81-229-XIB/81-229-XIB-e.html).

Warren, J.-P. 2009. "The Three Axes of Sociological Practice: The Case of French Quebec." *The Canadian Journal of Sociology/Cahiers Canadiens de Sociologie* 34(3):803–29.

Wente, M. 2012a. "Educated for Unemployment." *The Globe and Mail*, September 12. Retrieved February 2, 2016 (http://www.theglobeandmail.com/opinion/educated-for-unemployment/article4179127/).

Wente, M. 2012b. "Québec's University Students Are in for a Shock." *The Globe and Mail*, May 1. Retrieved February 2, 2016 (http://www.theglobeandmail.com/opinion/Québecs-university-students-are-in-for-a-shock/article4104304/).

Xie, Y. and K. Goyette. 2003. "Social Mobility and the Educational Choices of Asian Americans." *Social Science Research* 32:467–98.

Zarifa, D. 2012. "Choosing Fields in an Expansionary Era: Comparing Canada and the United States." *Research in Social Stratification and Mobility* 30(3):328–51.

JOBS, CAREERS & SOCIOLOGICAL SKILLS

The Early Employment Experiences of 2012 Sociology Majors

Mary S. Senter, Central Michigan University
Roberta Spalter-Roth, American Sociological Association and George Mason University
Nicole Van Vooren, American Sociological Association

While the United States has been officially out of recession since June, 2009, the unemployment rate hovers around six percent. Recent college graduates have experienced special difficulties, with suggestions that unemployment among this group exceeds eight percent (Shierholz, Davis, and Kimball, 2014). Further, student loan debt now is about one trillion dollars (Consumer Finance Protection Bureau, 2012), and the average debt per borrower from the Class of 2012 is $29,400 (The Project on Student Debt, 2013). In this environment, we would expect students contemplating a sociology major to ask what kinds of careers are common for people who have completed undergraduate degrees in the discipline. What kinds of jobs and what kind of career progress can graduates expect in their first post-baccalaureate years?

In this research brief, we will provide extensive detail on the jobs the national cohort of 2012 sociology graduates have obtained, along with information on the progress of their careers and their job satisfaction. Further, we will explore the kinds of skills learned in sociology programs that graduates find helpful on the job and how these contribute to job satisfaction, and the kinds of skills they wished they would have learned as undergraduates but did not. Elsewhere, we have examined the types of post-graduate jobs held by former sociology majors and whether particular types of jobs would have resulted in the desire to major in sociology again (Spalter-Roth, Van Vooren, and Senter 2014).

Data for this brief come from the 2012 Bachelor's and Beyond project funded by the National Science Foundation. The home page for the project, "Social Capital, Organizational Capital, and the Job Market for New Sociology Graduates," can be found at: http://www.asanet.org/research/bacc_survey/jobs_for_sociology_majors.cfm. For the first wave of the project administered in Spring,

2012, May and August, 2012 graduates from 160 colleges/universities were represented. Completed surveys were returned from 2,695 majors (37% response rate). The third wave of the survey was administered between November, 2013 and January, 2014, approximately 18 months after students' baccalaureate graduation. Third wave surveys were returned from 911 respondents.[1] The data are not weighted.

What Are They Doing?

This brief will, thus, focus on the 575 respondents whom we define as "workers." Fifty-six percent of the sample (N=504) reported that during the week of November 4, 2013 – the reference week – they were working at a paid job or internship only. Another 7.8 percent of the sample (N=71) indicated that they were both working and enrolled in a college or university, but that working was their primary activity.

Part-time Work

Almost 90 percent of those in the labor force reported that they held full-time positions, while about 12 percent of these respondents (11.6 percent) reported that their job was a temporary position, and 21.6 percent reported that they worked 34 hours or less per week. These 124 part-time workers were asked for the reasons for working fewer than 35 hours during a typical week on the job they held during the reference week. Answers were diverse with 32 stating that they did not need or want to work full time; 18 reporting that they were students; and 15 citing family responsibilities. Among respondents who provided an "other" response (which could be in addition to the responses summarized above), 56 simply said that the job was a part-time job or was limited to less than 40 hours per week, and only 23 reported that full-time work was not available.

Occupational Categories

Respondents were asked a number of questions about their specific job (regardless of whether their job was full or part-time). They were asked, first, "what category best describes the work you were doing during the week including November 4." Further, they were asked whether they perceived their job to be a career-type job. Table 1 provides the frequencies and percentages of respondents in each of eight self-selected major categories, along with the percentages of respondents who viewed their job as a career-type job.

At least 10 percent of the sample was found in each of four categories. The largest numbers of sociology graduates – more than 20 percent – were employed in social services or as counselors. About one sixth of sociology graduates (16.7 percent) were employed as administrative assistants or in clerical positions. An additional 12.6 percent of graduates worked in a variety of sales and marketing positions, and a similar percentage were employed as teachers (or, in a few cases, as librarians).

Career-Type Jobs

Overall, almost two-thirds (62 percent) of employees – those we have labelled as in the workforce – saw their job as a career-type job, while 24 percent said it was not. An additional 13 percent of respondents were not sure if their job was or was not a career-type job (data not shown). At least two thirds of respondents who held jobs within five of the major job categories reported that they are in career-type jobs (Table 1). Two of these job categories are among the four largest in terms of number of graduates occupying them – social services/counselors and teachers/librarians. The two categories whose job occupants were most likely to view their jobs as career-type – social science researcher and management-related job – include fewer respondents, although combined they comprise 13 percent of workers under analysis here. Another nine percent of workers were in other professional jobs, and three out of four of these respondents viewed their jobs as career type. Relatively few respondents considered their jobs in sales/marketing, administrative support/clerical, or service occupations to be career-type jobs.

---

[1] The second wave survey was administered between mid-December, 2012 and the end of January, 2013. The response rate for the second wave was just over 41 percent of those who had previously answered. Respondents who completed only the first wave or both the first and second wave surveys were invited to complete the third wave survey.

Job Descriptions

While the respondents' placement of their jobs into broad job categories is useful, they were also asked two open-ended questions – their job title and their description of "the kind of work you were doing during the week of November 4." Table 2 provides illustrative responses to these qualitative questions for the four largest job categories.

These descriptions suggest that the general categories suppress the variety and the levels of responsibility even in what respondents consider non-career jobs. The examples given here (e.g. by clerical workers) indicate that at

TABLE 1. Distribution of Respondents' Job Categories and Percent in Career-Type Jobs.

Job Category	N	% in Job Type	% in Career-type Job
Social Services/Counselors	123	21.9	77.9
Admin. Support/Clerical	94	16.7	42.4
Sales/Marketing	71	12.6	47.8
Teachers/Librarians	66	11.7	66.7
Service Occupations (waitresses, police, cooks)	52	9.3	26.5
Other Professional (inc. Prgm. Assistant, IT, PR)	50	8.9	75.5
Management-related	43	7.7	83.7
Social Science Researcher	30	5.3	90.0
Other	33	5.9	60.6
Total w/job category info.	562	100.0	62.2*
No info. on job category	13		

* Based on N=553

SOURCE: American Sociological Association. *Social Capital, Organizational Capital, and the Job Market for New Sociology Graduates, 2012, Wave III.*

TABLE 2. Examples of Job Titles and Work Descriptions by Major Job Category

Social Services/Counselors

Domestic Violence Victims Advocate

"I work as an advocate for victims of domestic violence in a public assistance office."

Residential Crisis Counselor

"Counselor for homeless and runaway youth in a shelter."

Justice and Peace Liaison

"I was hired by the diocese to heighten awareness of social justice issues at the parish level."

Teen Court Case Manager

"Case manager for first-time juvenile offenders participating in alternative program to traditional juvenile court."

Sales and Marketing

Account Executive

"I sell survey software – specifically I sell 360 feedback software."

Social Media/Marketing Analyst

"Assessed the performance of social media strategies as they relate to traffic growth, reader engagement, SEO, sales, and marketing. Created quantitative analyses using advanced Excel, Omniture, Google Analytics, and Wordpress."

Events and Marketing Coordinator

"I plan the events and do all marketing and communication for the mentoring and career services office..."

Marketing Associate

"Outreaching to hospital executives in order to schedule sales meetings for healthcare performance technologies."

Admin. Support/Clerical

Office Manager

"Budget management, general accounting, supervision of other employees."

HR Assistant

"Provide support within different parts of Human Resources (including recruiting, benefits, payroll, and director of HR)."

Executive Assistant and Externship Outreach Coordinator

"Assisting the CEO of the institution, along with assisting the Externship/Internship department."

Analyst

"Coordinating levels of management, facilitating communication between department, financial forecasting."

Teacher/Librarian

Elementary School Teacher

"I am a Kindergarten teacher: I instruct students in the Spanish language."

EFL Teacher

"I'm an English as a Foreign Language teacher in [Latin America] at a technical college..."

High School Special Education Teacher

"I am a high school special education teacher who supports approximately 120 10-12th students who receive special education services (...) in their English classes..."

Assistant Site Manager

"I manage a before and after school program for elementary students that includes activities, lessons and enrichment.

SOURCE: American Sociological Association. *Social Capital, Organizational Capital, and the Job Market for New Sociology Graduates, 2012, Wave III.*

least some respondents in non-career jobs have important responsibilities. Given the diversity of job categories, job titles, and job descriptions, it is not surprising that sociology graduates worked in a wide variety of employment settings. Table 3 provides a percentage distribution to summarize graduates' types of employers, along with the percentages of workers in each employment sector who considered their job a career-type job.

The largest numbers of sociology graduates (43.3 percent) were employed by private, for-profit companies. About one in four graduates (25.4 percent) worked for a private, not-for-profit, tax exempt, or charitable organization, and almost one quarter worked for the government at the local, state, or federal level. More than 70 percent of respondents who worked for a non-for-profit organization or for the government reported that their job was a career-type job as opposed to 54 percent of those who worked for for-profit companies. This finding suggests that for many for-profit companies might be employers of the last resort.

Twenty-five percent of all employed sociology graduates noted that their employer was an educational institution, with 42 percent of these 134 graduates indicating that their institution was a preschool or K-12 institution; 34 percent describing their employer as a 2- or 4-year university, college, technical institute, or a medical school

TABLE 3. Distribution of Employment Settings and Percent of Career-Type Jobs in these Settings

Employer Type	N	%	% in Career-type Job
Private for-profit company	242	43.3	53.6
Private non-profit, tax exempt, or charitable organization	142	25.4	73.2
Local govt, (e.g., city, county)	63	11.3	71.0
State government	56	10.0	75.0
Other	28	5.0	32.0
U.S. govt, as a civilian employee	12	2.1	66.7
Self-employed	11	2.0	36.4
U.S. military	5	0.9	100.0
Total	559	100.0	

SOURCE: American Sociological Association. Social Capital, Organizational Capital, and the Job Market for New Sociology Graduates, 2012, Wave III.

(including an affiliated hospital); 13 percent reported working for a university-affiliated research institute; and 11 percent indicating "other." In fact, 26 percent of graduates employed by private, not for-profit, tax exempt, or charitable organizations reported that their employer was an educational institution, along with 57 percent of graduates who worked for local government and 70 percent of those employed by state government. Looking at graduates with all types of employers, slightly more of those working at educational institutions (65%) reported their jobs as "career-level" compared to those working at non-educational institutions (60%).

Have They Seen Progress in Careers?

While these sociology graduates have had only so much time to progress in their careers, two sorts of questions provide insight into the ways in which this cohort of graduates have experienced career advancement. First, graduates were asked whether the job they held during the reference week was the only job they have had since graduating with their bachelor's degree in sociology. Forty two percent of the group said "yes." The remaining 328 respondents (58 percent) indicated all of the reasons they have changed jobs. Table 4 provides the percentages of these respondents who indicated each reason for the job change, recognizing that the percentages in the table will not equal 100 because respondents could indicate one or more reasons for changing jobs

The vast majority of respondents who changed jobs did so for a better job (71.3 percent). Job changes occasioned by layoffs were relatively uncommon, with about eight percent (7.9%) of those who changed jobs providing this as their only explanation for a job change or as one of multiple explanations. Regardless of whether respondents selected one reason or many, about one quarter of job changers (25.3%) reported that they relocated.

A second set of questions focused even more directly on job enhancement. Respondents were asked to indicate which of 11 positive changes they had experienced in their post-graduation employment (across all jobs). Table 5 provides the percentages of the 575 workers who had experienced each of these changes since graduation, with the most likely changes appearing first. The table also presents the percent of respondents experiencing and not experiencing each change who think of their job as a career-type job. We argue that respondents who have experienced job enhancement are more likely than those who have not to view their job as part of a career path. What is notable in the table is the high percentage of recent graduates who experienced positive changes. In fact, four types of job enhancements were enjoyed by more than one half of respondents – an increased level of responsibility, a salary increase, an increased degree of independence on the job, and an increased role in helping people.

To highlight further these job enhancements, we created an index by counting the total number of positive changes on the job (of 11) experienced by respondents. We found that 57 percent of respondents had experienced five or more of these 11 positive changes, and about one third experienced eight or more of them. Only 15 percent of respondents had seen none of these positive job

TABLE 4. Reasons for Job Changes, for Those Who Changed Jobs Since Graduation: Percentage Distributions

% for those who changed jobs: Combination of those who indicated one or multiple reasons for job change (N=328)

Better job	71.3 %
Relocation	25.3 %
Layoff	7.9 %
Other	25.9 %

SOURCE: American Sociological Association. Social Capital, Organizational Capital, and the Job Market for New Sociology Graduates, 2012, Wave III.

enhancements since graduation.

As seen in Table 5, respondents who reported that they have experienced positive change on the job are more likely than those who have not experienced job enhancement to view their job as a career-type job. At least 70 percent of those who experienced each of the 11 job enhancements reported that they held a career-type job. This compares to between 41 and 58 percent of respondents who did not experience the job enhancement who reported that they held a career-type job. Further,

TABLE 5. Experiences of Job Enhancement and Impact on Career-Type Jobs

% in All Jobs	%	% with Enhancement in Career-type Job	% without Enhancement in Career-type Job
Increased level of responsibility	64.0	73.0**	40.9
Salary increase	63.1	72.4**	42.9
Increased degree of independence	57.4	76.0**	42.0
Increased role in helping people	50.4	73.1**	50.2
More intellectual challenge	47.1	79.6**	45.6
Better opportunities for advancement	44.2	81.1**	46.2
Greater contribution to society	43.8	77.4**	49.5
Better co-workers	43.0	71.7*	54.6
Increase in job security	41.7	76.7**	51.1
Increase in benefits	39.5	75.2**	53.2
Better location	33.0	70.4*	58.0

* Statistically significant difference in the likelihood of being in a career-type job between those who experienced the job enhancement and those who did not at $p < .05$.

** Statistically significant difference in the likelihood of being in a career-type job between those who experienced the job enhancement and those who did not at $p < .01$.

SOURCE: American Sociological Association. Social Capital, Organizational Capital, and the Job Market for New Sociology Graduates, 2012, Wave III.

relatively few respondents who experienced each positive job change – in each case less than 20 percent – indicated that they held a non-career job, with the remaining 9-11 percent of respondents with each job enhancement reporting that they are not sure if their job was a career-type one (data not shown).

Are they satisfied on the job?

Next, we explore the factors that are associated with high overall job satisfaction. Table 6 provides data on the percent of graduates who are "very satisfied" overall with their job in each of eight major job categories and the category "other." Three other response categories were provided, with the most negative being "very dissatisfied."

Sociology graduates who worked in social services/counseling were most likely to report that they were very satisfied with their jobs. And, while the rank order differs, the five job categories with the highest percentage of respondents reporting career-type jobs (Table 1) are the five job categories with the highest percentages of very satisfied respondents. And, consistent with what we have reported above, sociology graduates in service occupations, sales/marketing, and administrative support/clerical jobs were least likely to be very satisfied with their jobs.

In addition to rating their level of overall job satisfaction, respondents were asked to indicate their satisfaction with 11 specific aspects of their job; these are the same 11 dimensions of job enhancements since graduation presented in Table 5. Table 7 is organized so that overall satisfaction is presented first followed by the specific aspects of the job with the highest "very satisfied" ratings in descending order. In addition, the table contrasts the satisfaction levels of those in career-type jobs with those not in career-type jobs, the logic being that those in career-type jobs are more likely than those in non-career jobs to be very satisfied overall and with specific aspects of their employment situation.

The table indicates that satisfaction levels of respondents were highest in terms of location, co-workers, degree of independence, and helping people. They were less satisfied with the material aspects of their jobs including benefits, opportunities for advancement, and salary. Overall, 37 percent of graduates were very satisfied and another 41 percent were somewhat satisfied with their jobs.

As expected, respondents in career-type jobs were more likely to be highly satisfied overall (52 percent) than respondents who were not in career-type jobs (7 percent). This general finding holds when one focuses on specific aspects of respondents' jobs. The distinctions are marked in some cases: 53 percent of those in a career-type job reported being very satisfied with their contribution to

TABLE 6. Percent "Very Satisfied" Overall with Job, by Job Category

Job Category	% "Very Satisfied" with Job
Social Services, Counselors	50.4
Management-related	45.2
Teachers/Librarians	37.5
Other professional (including program assistant, IT, PR)	37.5
Social Science Researcher	36.7
Other	31.2
Admin. Support/Clerical	30.8
Sales/Marketing	24.3
Service Occupations (waitresses, police, cooks)	23.9
Total	36.5

SOURCE: American Sociological Association. Social Capital, Organizational Capital, and the Job Market for New Sociology Graduates, 2012, Wave III.

society as opposed to only 11 percent of those not in a career-type job, and 59 percent of graduates in career-type jobs reported being very satisfied with their ability to help people in contrast to only 21 percent of those in other jobs. The relatively high percentages of graduates expressing high satisfaction with their ability to help people and to contribute to society is worth noting, given the high percentages of majors who noted in the first wave of the survey that they chose sociology majors for these reasons (Spalter-Roth, VanVooren, Kisielewski, and Senter 2012). Those in non-career jobs reported more satisfaction with specific items than with their job, overall. The satisfaction levels of the two groups of respondents were closest on salary.

TABLE 7. Overall and Specific Job Satisfaction: All Employees and for Those in and Not in Career-type Jobs

	% Very Satisfied	% Very Satisfied -- Those in Career Jobs	% Very Satisfied -- Those NOT in Career Jobs
Overall satisfaction	**36.5**	**52.1**	**6.8**
Location	55.9	62.7	43.5
Co-workers	49.5	56.8	38.2
Degree of independence	48.8	59.7	28.8
Helping people	47.4	58.5	20.6
Job security	42.5	51.3	25.0
Contribution to society	39.8	53.0	10.6
Level of responsibility	38.5	50.0	18.2
Benefits	34.6	45.5	16.0
Intellectual challenge	26.4	37.8	3.8
Opportunities for advancement	24.5	34.0	8.3
Salary	16.9	20.8	10.6

All relationships are statistically significant (chi-square test including the category "not sure" whether a career-type job, p=.000)

SOURCE: American Sociological Association. Social Capital, Organizational Capital, and the Job Market for New Sociology Graduates, 2012, Wave III.

Are They Using Their Skills on the Job?

Respondents were asked two sorts of questions about the extent to which they use concepts and skills learned in their undergraduate sociology program on the job. The first series of questions is whether or not they used basic sociological concepts and skills that are a central focus of the sociology curriculum (Tables 8 and 9). The second set of questions focuses on specific, practical skills presumably learned in their sociology courses (see Table 10).

Basic Concepts and Skills. Each statement in the series used this same format—in particular, "To help me in my job, I use what I've learned in my sociology undergraduate program about" each of nine specific concepts or skills and "other aspects." Elsewhere we examined the skills that people use on the job and whether they would major again (Spalter-Roth et al. 2014). Here, Table 8 provides percentage distributions to summarize responses

TABLE 8. Concepts and Skills Learned as Sociology Majors and Used on the Job: Percentage Distributions

To help me with my job, I use what I've learned about ...	Strongly agree	Somewhat agree	Somewhat disagree	Strongly Disagree
Diversity	55.2	28.1	7.4	9.3
Social institutions and their impact on individuals	49.9	31.1	9.4	9.6
Groups and teams	49.6	37.0	6.9	6.5
Social problems	47.5	26.4	12.9	13.1
Alt. or critical perspectives	41.1	31.1	15.2	12.6
Sociological imagination	33.8	34.9	14.7	16.5
Soc. concepts and theories	26.2	38.4	15.9	19.4
Data analysis	22.3	34.8	18.0	25.0
Other aspects	19.5	39.8	20.3	20.5
Research design	13.6	27.1	26.4	32.9

SOURCE: American Sociological Association. Social Capital, Organizational Capital, and the Job Market for New Sociology Graduates, 2012, Wave III.

to these questions, with the table organized so that the most used concept/skill appears first.

The table makes clear that many graduates strongly or somewhat agreed that the basic sociological concepts and skills that are part of the sociology curriculum helped them on their jobs. As we found with the relationship to whether graduates would major again (Spalter-Roth, Van Vooren and Senter 2014), sociological concepts appeared to be more useful than data analysis skills. The findings here are similar. Respondents were less likely to strongly agree that they use data analysis and research design on the job. Nonetheless, Table 8 shows that more than one half of respondents strongly or somewhat agreed that they used nine of the ten concepts/skills (with data

TABLE 9. Percent Very Satisfied with Job and Percent in Career-type Job by Extent of Agreement That Skills Learned in Undergraduate Program Are Helpful on the Job

	% VERY SATISFIED		% IN CAREER-TYPE JOB	
	Strongly agree skill helpful	Strongly disagree skill helpful	Strongly agree skill helpful	Strongly disagree skill helpful
Data Analysis	50.4	22.4	78.3	30.4
Research Design	49.3	21.8	82.2	34.7
Concepts and Theories	52.8	14.4	79.6	29.5
Sociological Imagination	55.0	18.0	75.8	29.2
Diversity	46.1	16.0	73.2	30.0
Groups and teams	46.8	11.8	73.4	28.6
Social institutions and their impact on individuals	47.8	11.5	75.9	28.8
Social problems	46.5	14.1	75.9	31.0
Alt. or critical perspectives	48.6	16.2	76.6	26.5
Other aspects	57.3	19.6	80.4	42.2

All relationships are statistically significant (chi-square test, p=.000).
SOURCE: American Sociological Association. Social Capital, Organizational Capital, and the Job Market for New Sociology Graduates, 2012, Wave III.

analysis and research design) under analysis here. In fact, about one half of graduates strongly agreed that they used what they have learned about diversity, about social institutions and their impact on individuals, about groups and teams, and about social problems to help them on their job.

But are respondents who find that basic sociological concepts and skills help them on the job more satisfied with their jobs than those who do not? And, are employees who find these skills helpful more likely to be in career-type jobs? Table 9 provides the data to answer these questions. The table presents the contrast – for each of 10 concepts/skills – between those who "strongly agree" and those who "strongly disagree" that "to help me in my job, I use what I've learned in my sociology undergraduate program."

The major conclusion from Table 9 is that respondents who strongly agreed that they used their sociological skills on the job were in every instance more likely to be very satisfied with their job than respondents who strongly disagreed that they use those skills. Similarly, respondents who strongly agreed that they use those skills on the job were in every case more likely to report that they are in a career-type job. For example, about one half of respondents who strongly agreed that they used data analysis learned as undergraduates to help them on the job were very satisfied with their job, and more than three quarters of respondents who strongly agreed that they used this skill were in career-type jobs. By contrast, only 22 percent of respondents who strongly disagreed that data analysis is helpful on the job were very satisfied, and only 30 percent who strongly disagreed that this skill is helpful were in career-type jobs.

Practical Skills. a second set of questions explored whether employees have specifically used practical sociological skills on their current jobs. These practical skills have the potential to link sociological analysis and methods to the heart of the job. Respondents were asked to indicate which ones of a set of skills "is something you have done on the job since you've had it." Table 10 provides those skills employed on the job by at least one third of respondents, arranged so that the activity most likely to have been done is listed first and the one least likely to have been done is last.

TABLE 10: Skills That are Part of the Job: Percentage Distributions

Skills	%
Organizational skills (including leadership)	85.9
Work with people who differ in race, ethnicity, gender or class	83.8
Work with others in teams	81.9
Use computer resources to locate information	69.4
Gather information to make evidence-based arguments	47.5
Summarize information in tables or graphs	42.3
Write a report	37.6
Make presentations using software such as PowerPoint	36.0
Search for existing statistics	35.1

SOURCE: American Sociological Association. Social Capital, Organizational Capital, and the Job Market for New Sociology Graduates, 2012, Wave III.

The percentage distributions themselves suggest three clusters of practical experiences. First, at least 80 percent of workers in the sample reported using the relationship skills that are the focus of major discussions in the sociology curriculum. These include organizational skills, working with people who differ in race, ethnicity, gender or class, and working in teams. Second, the empirical foundations of the discipline are realized on the job by the 42 to 69 percent of sociology graduates who reported having used computer resources to locate information, having gathered information to make evidence-based arguments, and having summarized information in tables or graphs and by the 35 percent of workers who have searched for existing statistics. Third, the importance of communications skills is highlighted by the 36-38 percent of respondents who reported having made presentations using software such as PowerPoint and having written a report.

What skills do they wish they had but do not?

What did graduates think that they were missing – the skills that real-world experience, post-graduation, led them to wish they had mastered as undergraduates? Former majors were queried about those skills that were not part of their undergraduate program that they think would have been useful. They were asked: "Given your experiences since graduating with your undergraduate degree, are there particular skills or topics that you wish you had learned while taking sociology courses, either for a job or for graduate school?" Three hundred and eleven "workers" provided an open-ended response to this question, which was coded into eight specific categories and "other." Table 11 provides the code categories, as well as the frequency and percentage distributions of responses based on N = 311, for those categories that include at least 10 percent of responses. Any one response could be coded into one or two categories.

The largest number of graduates who provided a response either wrote in the equivalent of "no additional skills" or provided a positive comment about their program. Examples of these comments when asked about particular skills or topics "that you wish you had learned" include:

I loved every part of the curriculum for getting a BA in Sociology.

No, I feel that my experience with my coursework leading up to my graduate degree prepared me academically and intellectually to be an effective and contributing member of society in a way where I feel as though I am making a difference.

No. I think the skills I learned during my undergrad career have helped me immensely in my post-college life. I have

TABLE 11: Skills They Wished They Would Have Learned: Code Categories, Frequency and Percentage Distribution

Code category	Number (1st and 2nd mention)	% (based on N=311)
Positive comments or no additional skills	83	26.7
Job information/internship/application	69	22.2
Specific course or content	61	19.6
Research/methods/statistics/statistical software	53	17.0

SOURCE: American Sociological Association. *Social Capital, Organizational Capital, and the Job Market for New Sociology Graduates,* 2012, Wave III.

learned that interpersonal, communication, presentation, and organizational skills are highly coveted by prospective employers - other role-based skills/knowledge can be learned on the job!

For respondents mentioning specific skills they wish they would have learned, the largest number focused on the need for more information related to jobs, including the desire for more internships or applied experiences. The following comments are illustrative of those included in this category.

How and where to find Sociology related careers.

I wish I was pushed harder by my teachers and advisors to do an internship and gain experience outside my college work.

I wish my adviser was better at connecting me with opportunities in the desired field I wish to enter.

Almost as many respondents mentioned a specific course that they wished they would have taken as undergraduates or mentioned a specific content area they wished they would have mastered. The range of these responses was wide and included:

Criminal justice-related skills

Critical race theory

Social work

Technology and society.

The only other category that includes more than 10 percent of the responses provided focuses on research, methods, and statistics. Graduates' comments coded in this category include:

I wish I perfected data analysis, statistics and research methods. I also wish I took more courses in economics.

Being able to use different computer programs for analyzing data, other than SPSS. Also, being able to conduct more research projects and actually getting out into the field and conducting more interviews and more ethnographic studies. Although I have been able to learn a lot in my courses, I would like for the program to be longer in order to spend more time on projects and be able to get more experience in such projects.

I wish I had learned a statistical software program. I also wish there was more focus on quantitative analysis rather than qualitative.

Note as well that additional comments did not explicitly mention statistical software (such as SPSS), but focused, more generally, on the desire for more computer skills (such as Excel). Perhaps in their job search or on the job they saw that a wider array of career-type opportunities were available to those with research and computer skills compared to the kinds of jobs that they had obtained.

Conclusions and Discussion

Eighteen months after graduation, sociology majors from the Class of 2012 were employed in a wide variety of occupations and employment settings. The largest numbers had jobs in social service and counseling, and almost 80 percent of graduates viewed this type of job as a career-type position. Further, one half of these graduates were "very satisfied" with their job.

Looking across all positions occupied by sociology graduates, we find that 62 percent viewed their positions as career-type jobs. While satisfaction levels were highest for those in career-type jobs, overall job satisfaction was strong for all employed respondents. Almost four out of five employees reported that they were very or somewhat satisfied with their position.

Sociology graduates found the concepts and skills they learned in their undergraduate programs were useful on the job. More than one half of respondents in all types of jobs strongly or somewhat agreed that – to help them on the job – they used nine of the ten concepts/skills (including "other") under analysis here. Sociology graduates who reported using their sociological toolkit on the job were also more likely to report that they were in career-type jobs and were more likely to report being very satisfied with their jobs.

Perhaps most important for the reputation of the sociology major, more than one half of sociology graduates reported having experienced four kinds of job enhancements since graduation, and – looking cumulatively across all changes – 57 percent of respondents have experienced five or more of these 11 positive changes including increased responsibility, more independence, greater intellectual challenges, more opportunities for advancement, and even raises in salaries. This evidence should be broadcast to improve the reputation of the sociology major.

These findings have implications for sociology departments and faculty. Faculty appear to have socialized undergraduates to love sociological concepts. It is important for faculty members to highlight the ways in which conceptual and methodological skills can be used effectively in the job market. Given that about one half of sociology majors are first-generation college students (Senter, Van Vooren, and Spalter-Roth, 2013), faculty members should not assume that the link between what is learned in the classroom and what is used on the job will be obvious for students.

Second, while sociology majors were employed in a wide variety of settings and many were satisfied with their jobs and have experienced advances in their careers, there remains a group of students who were dissatisfied with their employment and believed that their position would not lead them to where they want to be in the next five years. For these students in particular, but also for the benefit of all students, departments need to help link majors with career resources. Previously we have shown that less than 30 percent of undergraduate majors were satisfied with the career advising that they obtained (Senter, Van Vooren, and Spalter-Roth 2013). To improve the level of satisfaction, departments could cultivate links to the career services units on their campus and link their department or program websites to those of departments that have been especially successful in creating career-oriented sites for undergraduates (see the appendix of Spalter-Roth, Van Vooren, and Senter 2013 for suggestions). Further, while the data provided here are from a national sample of majors, departments should consider maintaining links to their own alumni in a variety of ways. Alumni surveys (perhaps using some of the questions from this project and available for download at: http://www.asanet.org/research/bacc_survey/jobs_for_sociology_majors.cfm) will provide data on what a department's particular graduates are doing; such data might prove especially compelling to your students and their parents. Further, by maintaining ties to your alumni, you may find potential speakers and panelists for events that you wish to host for your current undergraduate majors.

Third, departments should consider the ways in which their undergraduate curriculum might best serve the career needs of students, without in any way detracting from sociological substance and rigor. When asked about what was missing in their undergraduate program, the largest number of employed respondents in our sample talked about the need for more information related to jobs, including the desire for more internships or applied experiences. Department faculty might want to discuss whether an internship program or a pro-seminar could be added to the curriculum, or whether a capstone course could devote (more) attention to career issues. Further, more than one in six respondents who articulated additional skills that they wished they had learned focused on research methods, data analysis, and statistics. More than 40 percent of all employed sociology graduates told us that their jobs have involved using computer resources to locate information, gathering information to make evidence-based arguments, and summarizing information in tables or graphs. While a relatively small number of graduates worked as social science researchers so early in their careers, the kinds of skills learned in methods and statistics courses proved valuable to students-as-employees. This information, communicated to today's undergraduates, might serve to motivate them to take such courses seriously and to do well in them.

References

Consumer Finance Protection Bureau. March 21, 2012. "Too Big to Fail: Student debt hits a trillion." Retrieved on August 11, 2014 from http://www.cosumerfinance.gov/blog/too-big-to-fait-student-debt-hits-a-trillion/

National Association of Colleges and Employers. September 4, 2013. "Salary Survey: Average Starting Salary for Class of 2013 Grads Increases by 2.4 Percent." Retrieved on December 20, 2014 from https://www.naceweb.org/s09042013/salary-survey-average-starting-class-2013.aspx.

Payscale.com. "2012-2013 Payscale College Salary Report." Retrieved on December 20, 2014 from http://www.

payscale.com/college-salary-report-2013/majors-that-pay-you-back.

Shierholz, Heidi, Alyssa Davis, and Will Kimball. May 1, 2014. "The Class of 2014: The Weak Economy Is Idling Too Many Young Graduates." EPI Briefing Paper #377. Washington, DC: Economic Policy Institute.

The Project on Student Debt. December 4, 2013. "Average Student Debt Climbing: $29,400 for Class of 2012." Retrieved on August 11, 2014 from http://projectonstudentdebt.org/

Research Briefs from "Social Capital, Organizational Capital, and the Job Market for New Sociology Graduates"

Spalter-Roth, Roberta, Nicole Van Vooren, and Mary S. Senter. 2014. "Recent Sociology Alumni: Would they Major Again?" http://www.asanet.org/documents/research/pdfs/Bach_Beyond_MajorAgain.pdf Washington, DC: American Sociological Association.

Spalter-Roth, Roberta, Nicole Van Vooren, Michael Kisielewski, and Mary S. Senter. July, 2013. "Strong Ties, Weak Ties, or No Ties: What Helped Sociology Majors Find Career-Level Jobs." http://www.asanet.org/documents/research/pdfs/Bach_Beyond5_Social_Capital.pdf Washington, DC: American Sociological Association.

Spalter-Roth, Roberta, Nicole Van Vorren, and Mary S. Senter. March, 2013. "Social Capital for Sociology Majors: Applied Activities and Peer Networks." http://www.asanet.org/documents/research/pdfs/Bach_Beyond4_Social_Capital.pdf Washington, DC: American Sociological Association.

Senter, Mary S., Nicole Van Vooren, and Roberta Spalter-Roth. January, 2013. "Sociology Majors: Before Graduation in 2012." http://www.asanet.org/documents/research/pdfs/BachBeyond_Sociology_Majors_Brief_2013.pdf Washington, DC: American Sociological Association.

Senter, Mary S., Nicole Van Vooren, Michael Kisielewski, and Roberta Spalter-Roth. November, 2012. "What Leads to Student Satisfaction with Sociology Programs?" http://www.asanet.org/documents/research/pdfs/Bachelors_and_Beyond_2012_Brief1_Satisfaction.pdf Washington, DC: American Sociological Association.

Spalter-Roth, Roberta, Nicole Van Vooren, Michael Kisielewski, and Mary S. Senter. November, 2012. "Recruiting Sociology Majors: What Are the Effects of the Great Recession? Concepts, Change, and Careers." http://www.asanet.org/documents/research/pdfs/Bachelors_and_Beyond_2012_Brief2_Recruiting.pdf Washington, DC: American Sociological Association.

All ASA research briefs are available for free online at:
http://www.asanet.org/research/briefs_and_articles.cfm

American Sociological Association

1430 K Street, NW Suite 600

Washington, DC 20005

www.asanet.org

HUMAN ARCHITECTURE: JOURNAL OF THE SOCIOLOGY OF SELF-KNOWLEDGE
A Publication of OKCIR: The Omar Khayyam Center for Integrative Research in Utopia, Mysticism, and Science (Utopystics)
ISSN: 1540-5699. © Copyright by Ahead Publishing House (imprint: Okcir Press). All Rights Reserved.

Choosing My Major and Career
A Sociological Inquiry

Jacquelyn Knoblock

University of Massachusetts Boston

jacquelyn.knobloc001@umb.edu

Abstract: In this paper, applying various micro- and macrosociological theories and concepts, I explore how I ended up taking Social Psychology as a major and what my options are for a future career. I use C. Wright Mills's notion of the sociological imagination as a way of looking at and interpreting the circumstances of my life and my feelings and reactions towards them. I explore how such circumstances and explorations have caused me to be where I am, and how they may influence where I am going. I find that it was mostly personal troubles that led me to this field, but it is public issues that keep me interested and that make me want to continue in this field as a profession. I understand the tension of opposites between what I want to do with my life and my time, and what I must do to "make it" in this society. I am coming to see that society's view of success and my own may differ and they don't have to be the same. I see that my learning of different sociological theories and perspectives has enriched my own viewpoints and that I desire to be able to extend these viewpoints to others through teaching and/or counseling, and that these fields would seem to suit me because I am comfortable in the realm of academia. I do not want to get stuck in the "rat race" of corporations and capitalism, but that I must do something to make money to be able to support myself and live a basic comfortable life. Therefore I must make a compromise between the opposites. I realize that it is my personal fears and insecurities as well as some of the larger and public institutions that do not make it easy for women to succeed both in the workplace and in the home. These may hold me back from making high goals for myself and following through. However, I also see that I should not worry too much about getting a job and what I am going to do, and should instead focus now on learning about what interests me in the hopes of finding a meaningful occupation.

My current major is Social Psychology. This means that half of my classes are under the sociology department and the other half under the psychology department. When I tell people about my major, the usual reply is "Oh, that's interesting, what are you planning on doing with it?" My response to this is always, "I'm going to go to Grad School." Then I'll usually give a list of a few fields of graduate studies that I am interest in, and four or five things that I may end up doing with my possible future degree. I can't limit myself to one choice, and I think I am afraid of choosing a goal, for fear of not achieving it.

I used to think that Sociology and Psychology were both just phony majors, full of information that was just commonsense,

---

Jacquelyn Knoblock is an undergraduate junior at UMass Boston, majoring in Social Psychology. She wrote this paper while enrolled in the course Soc. 341-1: "Elements of Sociological Theory," instructed by Mohammad H. Tamdgidi (Assistant Professor of Sociology at UMass Boston) during the Fall 2007 semester.

and for people who were bleeding hearts or who didn't know what they wanted to major in. Sociology and/or Psychology were never on my possible lists of majors during my senior year of high school. It was a difficult road that led me to this major, with the most salient events occurring in the past two years, but on a road that probably began as a young child.

In this paper, using a framework informed by various sociological theories, I am going to investigate what caused me to choose Social Psychology as a major and explore the possibilities for my future career goals. In order to do this I must employ what C. Wright Mills has termed the **Sociological Imagination**. By this notion, Mills means the ability to relate one's own life with society as a whole, and the capacity to shift views from the personal to the public (348).

To understand my choice in becoming a Social Psychology student and to determine my future educational goals, I will first look at what Mills has termed **"troubles."** Troubles happen "within the character of the individual and within the range of his immediate relations with others" (Mills 350). Troubles fall under the realm of **microsociology**, which is concerned with face-to-face encounters and the details of communication and human interaction (Wallace & Wolf 5). It is the microsociological connection that I will focus on during the first half of my paper.

To begin to question and comprehend the reasons I have chosen Social Psychology and what I plan to do with my degree, I must first **problematize** it from a **stranger's** point of view, and look at my life as though I have no preconceptions of what it should be, suspending my "learned cultural notions" (Wallace and Wolf 262). These approaches were used by Harold Garfinkel, the founder of **Ethnomethodology,** as ways of explaining his methods. Ethnomethodology is people's ways of making sense of their social world (Wallace & Wolf, 269). Ethnomethodology is a type of research associated with **Phenomenology.** Phenomenology is interested in things that can be directly understood by one's senses (Wallace & Wolf 263) and "asks us not to take the notions we have learned for granted, but…to question our way of looking at and our way of being in the world" (Wallace & Wolf 262). I need to look at my past experiences and problematize and understand them to question the place I am in today, and to figure out where I want to go in the future. I agree with the UMass Boston student Jennifer Kosmas that "If we can become aware of our own social constructions, then we can also break out of the self-destructive habits by knowing ourselves" (74). So if I gain awareness of what has led me here, I can begin to know myself and figure out the best path for my future.

In her article titled, "The Roots of Procrastination: A Sociological Inquiry into Why I Wait Until Tomorrow," Kosmas analyzed her relationship with her father in order to get to the bottom of her procrastination problem. Similarly, in her essay titled, "Accepting Myself: Negotiating Self-Esteem and Conformity in Light of Sociological Theories," Sheerin Hosseini discussed her relationship with her parents in terms of her issues dealing with self-esteem. I feel that both procrastination and self-esteem are issues for me as well. I try to avoid any type of conflict that will make me feel uncomfortable, and if I have an assignment that seems difficult or that I am not sure of, I will postpone it, perhaps in an effort to self-handicap.

This may also stem from my relationship with my parents, and my place in my family. The family is the **primary group** of **socialization**. This means that individuals in childhood come to identify emotionally with the significant others in their lives, and that they come to **internalize social norms** and **values** (Wallace & Wolf 290). My family was not rich, but we lived comfortably, and my parents did not force us to do chores or even to really clean up after ourselves very much. We were free to enjoy our childhood, which was nice at the time, but I think I may have overly internalized the role of a child, and that now this is why I have a fear of growing up, and can't imagine living on my own, and having all the extra responsibilities of paying my bills, cooking dinner, and cleaning up. My father is also what may be called a "helicopter dad" because he tends to hover over our affairs even now, so I feel like I have internalized a sense of helplessness in dealing with things that my dad usually takes care of. I think that part of the reason I may have chosen Social Psychology as a major is because it would take me a few extra years, since not many of the credits from my former major and my previous colleges transferred into this major. Also, I know that I will probably have to attend graduate school. Overall, these give me more years of schooling where I do not have to be a full "grown up."

Another reason that I have chosen to continue my education and become a social psychology major is because being a college student is important to me. In my family my parents did value education, and there was never any question that we could all go to college. When I started elementary school I moved into **secondary socialization**. Secondary socialization concerns more specific **roles** and is a later phase of socialization of an individual who has already been socialized (Wallace and Wolf 290). In these classes, I took on the role of the "brain" and sometimes the "teacher's pet." In her essay cited previously, Sheerin Hosseini states that, "As a child, it seemed I wanted more than anything to be noticed and admired by both my parents and my teachers. I wanted to be the one in class who got recognition from the teacher for being intelligent or artistic" (29). I felt very much the same way as Hosseini. Once I realized that I was good at learning, that it came easier to me than most of my classmates, I ran with it, and took this as an important part of my self-concept and self-esteem. This has led me to achieve, but also has given me an extreme fear of failure. I feel like educational success is such an important part of my self-concept that I generally avoid trying activities in which I may not succeed, or that I feel might be too hard, and that I won't fare well in.

I have taken my role as a student seriously for the most part. In the film *Twelve Angry Men*, the jurors come in with their own presupposed ideas of what a jury was supposed to do and what the role of a juror was. Some took the roles more seriously than others. They created their own arbitrary rules and followed them. This also connects to the work that Garfinkel did on jury deliberation. The jurors "knew how they were supposed to act" by drawing on knowledge from other sources (Wallace and Wolf 276). In the same way I, even as I entered elementary school for the first time, or entered college as a freshman, knew what was expected of me by drawing on knowledge from other sources, what I had learned from books, TV, movies, friends, and parents.

It's also important for me to be viewed by others as a college student. I work as a cashier, a job that isn't very prestigious, and I always feel the need to make it known that my first priority is being a student, and that I only work part time to pay tuition and have some spending money. I also feel that some of my reasons for wanting to earn a college degree are so that I can live "the good life." The film *Affluenza* illustrates how Americans are consumerists and that many of us equate happiness with spending, and the goods that we acquire. Although I don't feel that making money is the greatest factor in my decision to major in Social Psychology, I can't deny that the prospect of money is an important aspect of my choice to attend college.

The most significant event in my life that caused me to think about work in Social Services or Psychology and that led me to become a Social Psychology major was being involved in an abusive relationship. I became involved with a man who at first seemed like a good boyfriend, then he became controlling, emotionally abusive and finally physically abusive. It was during this time that I withdrew from my first college. I can analyze this time in my life by drawing on many of the concepts included in **Symbolic Interactionism** and **Rational Choice or Social Exchange Theory**.

Symbolic Interactionism is "a social-psychological perspective; it focuses primarily on the individual "with a self" and on the interaction between a person's internal thoughts and emotions and his or her behavior" (Wallace & Wolf 199). Rational Choice Theory assumes that people are rational and make their decisions and perform actions based on what they think are the most effective means to their goals. The theory presumes that people are constantly weighing alternatives.

There are many concepts in these theories that apply to my situation. At first I allowed a little bit of controlling in the relationship, because of self-esteem issues and my problems with making my own decisions, and because I enjoyed the **status** that came with having a boyfriend. In Social Exchange Theory, status is defined as common recognition by others of the amount of esteem or friendship that someone receives (Wallace and Wolf 340). Once my troubles got more serious, however, I tried to hide them, and refused to admit to anyone that anything was wrong. I engaged in what Goffman terms **impression management**; I tried to control how others viewed me. That I could do this may be explained best by Goffman's theory of **dramaturgy**. Dramaturgy is concerned with the lives of ordinary people, and how they act out their daily roles on the "stage" of life (Wallace & Wolf, 239). On the **front stage**, in the presence of an audience, I acted as if everything was perfectly fine, that I had it all together, and refused to let on that anything was wrong. It was only in the **back region**, hidden away from others, that any of the abuse took place and that I could cry and let myself feel the pain.

During this time, I felt like I was a **discreditable person**, that I carried this **stigma** of being an abused woman which could be discovered at any time. It seemed to me, if I didn't let anyone else know that something was wrong, then nothing was wrong; but if anyone else found out, I couldn't go on pretending anymore. Finally, when it became more than I could handle with a smile or an explanation, I began to isolate myself from my friends and family. I even lost my sense of **self**. George Herbert Mead describes the self not as a "passive receptacle" but as active and creative (Wallace & Wolf, 205). At this point in my life, I was not active and creative, I only felt that I was the "**me**" and not the "**I**," that I was acted upon, but not acting or reacting upon my own initiative. I left my first college at this time. Eventually my parents made me go to therapy (once again, I did not make my own decision). I ended up finally getting out of the relationship and moving to South Carolina.

It was in South Carolina that I decided that I wanted to take some classes in Psychology and Sociology to try to understand both myself and others better, and hopefully see what had happened to me. In this sense, I was sort of like Neo, the main character in the movie *The Matrix*. In one scene, Neo is given a choice whether to take a pill and gain an understanding of what the real world is, without any promises that such knowledge would make him happy, or that he could go back to living his artificially created life. Neo decided to take the pill that would cause him to learn about what was really happening. My choice was similar; I wanted to gain an understanding of myself and other people and the structures of society. I felt that maybe I could get some answers by studying Social Psychology. As Avery Gordon wrote in her book *Keeping Good Time*, "faced with our powerlessness to assert authoritatively what is going on, we only want to 'simply understand'" (9). Gordon was speaking about research during wartime, but I feel that her statement applies to most people who can't ascertain what is happening in their lives or feel powerless to take action to change it, but at least want to understand what is going on.

Now that I have gained some understanding, the question I have is about where I want to go once I get the degree. I have basically determined that graduate school will be necessary. But, I suffer from an anxiety that I won't be accepted, and I know I probably won't apply to any elite schools, because of the fear of being rejected. I have never been rejected by a college, and I think that since so much of my identity is tied up in my academic abilities, that rejection would hurt my sense of self. I limit myself to schools I know I will probably get into. I also limit myself in terms of being a woman and knowing that if I have children, I will want to have a job in which I can fit my family life and career.

In her essay titled "Altruism or Guilt: Applying My Sociological Imagination to Choosing a Helping Profession," Elizabeth McCauley writes about how her future career choices are based on ones that would have a mother-friendly schedule; but she also states that, "I am not certain I want to have children. I am positive that if I choose to I will be in my late thirties when I start, so why am I basing my career choices on nonpriorities or priorities that are not my own?" (149).

Similarly, I am worried about experiencing the **bifurcation of consciousness**. This is experienced in women as they move between modes, generally work mode and home mode, and are constantly aware of and concerned about each, because there are ongoing pressing demands on women, especially those with children (Farganis, 375). Therefore, when I think about wanting to be a professor, I usually think about just being an adjunct so that I would still have time to have a family. Or, I may think I want to have my own counseling practice, but then I'll think that may take up too much time, or be too hard. Because of the expectation of women to raise a family, I think about it even when I am in no position to be having a family anytime in the near future. I do think that I want to help or at least educate others to avoid, or get out of the situation I was in when I was involved in the abusive relationship. But I struggle with this: is it this because I really want to help others, or is it just because I want to give meaning to this awful thing that I went through, the powerlessness that I felt? Avery Gordon states that,

> The struggle to achieve a certain independence from the ways we are imprisoned, with love in our hearts, preparing for a struggle that will change us, that will make us more free, more capable of sharing freedom than we were before: that's *the place we go to, the way we get there*. (66)

I feel like I have gone through a struggle, and part of the reason I chose Social Psychology was for understanding, and the other part was so that maybe in the future I would help others by sharing what happened, or just by being someone who could listen to their situations and not judge them. This is where I feel that I want to go, and I think that it is only my own apprehensions, and my interpretation of society's ideals, that can hold me back.

In his book, *The Sociological Imagination*, Mills wrote that distinguishing personal troubles—which I defined and explored in the first half of my paper—from ***public issues***" is an essential tool of the sociological imagination" (p.350). Public issues have to do with matters that involve the larger social structures and impact society as a whole. These are part of the realm of ***macrosociology***. Macrosociological perspectives are concerned with analyzing "the large scale characteristics of social structures and roles" (Wallace & Wolf, p 5). Macrosociological theories include ***Functionalism***, ***Conflict Theory***, and ***Postmodernism***.

The functionalist perspective states that society is made up of different parts and that these parts must work together in order for society to function as a whole and maintain equilibrium (Wallace & Wolf 16). Conflict theory assumes that people have certain basic needs and wants, that power is at the core of social relationships, and that values and ideas are seen as weapons (Wallace & Wolf, 68). Postmodernism has come to mean different things, it has been associated with post-industrialism and post-Marxism, a movement in literary criticism. It deals with the Enlightenment promise of rationality and the relationship of knowledge to theory and power (Farganis 423).

We are influenced by the macrostructures of our society. In our socialization we are introduced to the prevailing ideas of our culture. Culture is defined as "the enduring behaviors, ideas, attitudes, and traditions shared by a large group of people and transmitted from one generation to the next" (Myers, 11). This transmission of culture through generations occurs through socialization and is part of what Emile Durkheim's called ***integration***, "the incorporation of individuals into the social order" (Wallace & Wolf, 20). Durkheim maintains that integration is important for the maintenance of social equilibrium.

An important part of integration is the understanding of the current ideologies comprising and shaping the culture. Conflict theorists, such as Karl Marx regarded established ***ideologies*** as beliefs that serve to legitimize the position of those who currently are in control (Wallace & Wolf, 87). I will discuss some of these ideas that have been internalized by most people and myself, and how they affect our perceptions of the world, and what we end up doing with our lives in it.

The contemporary U.S. society has an ideology that admires rugged-individualism. In our individualistic society identity is self-contained, becoming self-reliant is viewed as an important trait of a successful individual, and independence is valued (Myers, 41). As the division of labor increased in our society so did individualism, or at least the idea of it.

I have struggled with this individualism in my life. I have always been a dependent person; I haven't really lived on my own. I think that I want to at some point (or maybe just because society wants me to), but I don't have the money, and I think I am also scared of the prospect of not having my family or friends nearby for support. This will also limit my prospects for the graduate school and careers as I will probably stay in the Boston area or the South Shore. In the movie and book *Tuesdays with Morrie* by Mitch Albom, the late sociology professor from Brandeis Morrie Schwartz says that our culture tells us we should be ashamed of being dependent on others, but that "there is nothing innately shameful about being dependent." Most people in the U.S. would probably disagree with Morrie. I struggle with this because as I grew up I was allowed to be dependent, and never learned to do things for myself; but now at my age that dependency is deemed inappropriate, and I am told that I need to just grow up. Our society values independence and abhors dependency so much that we have pathologized dependency as a personality disorder in the DSM-IV.

The paradox is that we are not truly independent in our society; we are interdependent. We rely on what Durkheim termed ***organic solidarity***, which occurred as the division of labor and ***specialization*** increased. Organic Solidarity is "characterized by the interdependence of roles and a lack of self-sufficiency" (Wallace & Wolf, 20). Durkheim states that this is what holds people together when they are all performing separate tasks. Ferdinand Toennies first termed this type of modern industrial society a ***gesellschaft***, which is "characterized by a predominance of more impersonal or business-type relationships" (Wallace & Wolf, 30). The Functionalist Talcott Parsons labeled these types of relationships in the modern society ***instrumental***. The focus on independence, these formal relationships, as well as the exploitative relationships of ***Capitalism***, are what allow most people to justify their own or others' positions in life. They feel that whether one is on top of the chain or on the bottom is a product of their own hard-work. I used to feel this way, and still in part hold on to this ideology for myself; I think that if I just work hard enough and find my niche I will be able to support myself, make money, and live the "good-life." Of course there are many other factors that influence these outcomes other than "hard-work."

Although I still hold on to this ideology for myself, I do not expect it of others. A concern that has brought me into the Social Psychology field of study is that of the welfare of others. At my previous college I took business, accounting, management and economic classes. During these classes I never felt good about profits. I felt that leadership and management classes were all about manipulating people to get the most productive value out of them. In the book and movie *Tuesdays with Morrie*, Morrie was brought by his father to the factory where he worked making fur coats. Having seen and experienced the inside of the factory Morrie vowed that he would, "never do work that used people, that made money off their sweat and degraded them." I also felt this way; how could I be concerned about business and profits and not worry that the people in the company weren't happy or were being exploited or were in danger of losing their jobs?

Michael Moore tackles this issue in his film *The Big One*, where he illustrates some of the real casualties of capitalism. During one stop on his book tour, Moore visits Centralia, Illinois, home of the PayDay candy bar. When he arrives he finds that the workers in the PayDay plant just found out (through a video) that they were losing their jobs. The ironic part was that the workers had worked hard, done their jobs well, and made profits, **surplus value**, for the company. The factory workers were termed **proletarians** by Marx, the laborers who don't own any of the means of production. These particular workers in Centralia are the embodiment of modern **exploitation**. Karl Marx described exploitation as occurring when "any surplus value is appropriated by someone other than the worker" (Wallace & Wolf 84). Because of this exploitation, the company had the means to relocate their factory to another country with a cheaper labor force.

Of course at the PayDay factory in Centralia, and in many of the other corporations Moore visited during his book tour, the management and the owners and CEO (the **bourgeoisie**, the people who "control capital and direct labor" [Wallace & Wolf 90]) did not seem to care about the people they had let go. To them it was all about "remaining competitive" and "having an obligation to the stockholders." Many of the CEOs had a corporate mentality in that in their thinking and conversation they became a spokesperson for the corporation and had the interests of the corporate structure in mind. This corporate mentality included what Max Weber called **Zweckrational**, a "form of reasoning that breaks down all problems into a means-ends chain and entails rational calculation of costs incurred and benefits to be secured if a particular line of action is pursued" (Wallace & Wolf 93). The management and owners dis not seem to worry or care about what happened to their employees, because in the end they would be making profit by laying them off and downsizing to another country. This is a feature of capitalistic mentality and *Zweckrational*.

Weber came up with the term the latter term to describe **bureaucracies**. Features of bureaucracies include highly specialized tasks that are arranged in a hierarchical order, depersonalization, routinization, and mechanical predictability. I do not want to end up working in a big bureaucracy or dealing with the **alienation** and impersonalization of a job like that. Therefore I am caught in what Morrie Schwartz termed, "a tension of opposites" and what Karl Marx termed **dialectics**, a contradiction, or a "pattern of inner conflict" (Wallace & Wolf, 162). I, like most people, am stuck between what I want to do and what I must do. My dream would be to be a perpetual student, and I wish I could just learn for the sake of learning and not deal with the bureaucracy of deadlines and grades, etc. I enjoy learning and my ideal would be to be learn about everything and anything that I find interesting. Unfortunately, that is not the "real-world." I have already been in college for 5 years and can't afford to continue indefinitely. I need to finish college and earn a degree so that I can get a "real job."

Also the problem with school for me is that I am so concerned about grades and I really don't learn for the sake of learning. I wish I could, but I always know that the grade is what is really important. That is the way I can maintain scholarships. Graduate schools and future employers are going to judge me chiefly by my GPA.

Although grades are important to me, I hope to become more like the alternative students that Avery Gordon speaks about:

> By naming yourselves alternative students you have chosen a path, a perspective, that says you are part of a long tradition of those who see differently, who imagine and dream other possibilities than the normal ones taken for granted. You have aligned with those…who see the individual as part of the larger social fabric and consider this co-operative interaction freedom; with those who feel a source of pride and accomplishment in what is derided and undervalued; with those who don't take the easy way out but are willing to live and struggle with difficult contradictions. (72)

One of the reasons I decided to attend UMass Boston was because it is more of an alternative campus. There are students of all ages, from many different ethnicities, and **socioeconomic statuses**. There are people who go to school full-time, and those who work full-time and take one or two classes a semester. I wanted to be exposed to a wider range of perspectives, which I get from the students around me and from being in a major that really consists of two majors.

I feel that the more I learn, the more I want to learn. I definitely feel that in the future I would like to counsel or teach. I think that it is important for people to be exposed to viewpoints that take them out of the conventional ways of thinking about the world. Rose Weitz writes about resistance in her article, "Women and Their Hair: Seeking Power Through Resistance and Accommodation." She states that we must define resistance as "actions that not only reject subordination *but do so by challenging the ideologies that support that subordination*" (137). In order for people to challenge and resist the social forces that act on their lives, they must first understand the ideologies that have put the forces in place and where they come from, to be able to think of different ways of thinking or living. By teaching or counseling I feel that I would have the means to at least expose people to some differing paradigms or ways of looking at life in hopes of allowing them opportunities to resist. Avery Gordon writes of this as dissent:

> The maintenance of an alternative perspective requires also an active commitment to dissent and to the insistence that dissent is never treason … to commit dissent is to commit self-governance, not administrative governance … social justice, not personal justice … to commit to learning and unlearning again and again, with pleasure. Dissent is the key link between education and democracy. (73)

I see that this dissent or resistance is important, and that education is a key factor in understanding systems and having the ability to resist them. I myself have always been a firm believer in learning and empiricism and the scientific method, but in reading about the Postmodern perspective I see that there is some room for subjective relativity concerning reality and that knowledge and science are bound with power. I feel that understanding this has made me less dogmatic in my views, and I feel that interaction with others on a face to face or a teaching level would continue to open my life towards new perspectives.

To conclude, while trying to understand what has brought me into the field of Social Psychology and what I am going to do with my life when I finish school, I see that it was mostly personal troubles that led me to this field, but it is public issues that keep me interested and that make me want to continue in this field as a profession. I understand the tension of opposites between what I want to do with my life and my time, and what I must do to "make it" in this society. I am coming to see that society's view of success and my own may differ and they don't have to be the same. I have seen that my learning of different sociological theories and perspectives has enriched my own viewpoints and that I desire to be able to extend these viewpoints to others through teaching and/or counseling, and that these fields would seem to suit me because I am comfortable in the realm of academia.

I do not want to get stuck in the "rat race" of corporations and capitalism, but that I must do something to make money to be able to support myself and live a basic comfortable life. Therefore I must make a compromise between the opposites. I realize that it is my personal fears and insecurities as well as some of the larger and public institutions that do not make it easy for women to succeed both in the workplace and in the home. These may hold me back from making high goals for myself and following through. However, I also see that I should not worry too much about getting a job and what I am going to do, and should instead focus now on learning about what interests me in the hopes of finding a meaningful occupation.

REFERENCES

Albom, Mitch. *Tuesdays with Morrie.* New York: Random House, 1997.

Farganis, James (ed.) *Readings in Social Theory: The Classical Tradition to Post-Modernism.* Fourth Edition. McGraw Hill College Division, 2004.

Gordon, Avery. *Keeping Good Time: Reflections on Knowledge, Power and People,* Paradigm Publishers, 2004.

Hosseini, Sheerin. "Accepting Myself: Negotiating Self-Esteem and Conformity in Light Of Sociological Theories." *Human Architecture: Journal of the Sociology Of Self Knowledge*, IV, 1&2 Fall 2005-Spring 2006, 29-44

Kosmas, Jennifer. "The Roots of Procrastination: A Sociological Inquiry into Why I Wait Until Tomorrow." *Human Architecture: Journal of the Sociology Of Self Knowledge.* Vol. II, No. 2, Fall 3003-Spring 2004. 74-81

McCauley, Elizabeth. "Altruism or Guilt: Applying My Sociological Imagination to Choosing a Helping Profession." *Human Architecture: Journal of the Sociology Of Self Knowledge*, IV, 1&2 Fall 2005-Spring 2006, 147-156.

Mills, C. Wright. The Sociological Imagination. Oxford: Oxford University Press, 2000.

Myers, David G. *Social Psychology.* Ninth Edition. McGraw Hill, 2008.

Wallace, Ruth A., and Alison Wolf. *Contemporary Sociological Theory: Expanding the Classical Tradition,* Sixth Edition. New Jersey: Prentice Hall, 2005.

Weitz, Rose. "Women and Their Hair: Seeking Power Through Resistance and Accommodation." *The Politics of Women's Bodies: Sexuality, Appearance, and Behavior.* Second Edition. Ed. Rose Weitz. New York: Oxford University Press: 2003.

Films:

Affluenza. Bullfrog Films, 1997.

The Big One. Miramax Home Entert., 1997.

The Matrix. Warner Brothers, 1999.

Tuesdays with Morrie. Touchstone, 1999.

Twelve Angry Men. MGM, 1957.

CALLED TO ACCOUNT: THE CV AS AN AUTOBIOGRAPHICAL PRACTICE

NOD MILLER AND DAVID MORGAN

Abstract The curriculum vitae has become a central feature of modern academic life. It can be understood as a form of autobiographical practice, one centrally involved with the construction and presentation of a self in a particular occupational context. This paper explores these themes and seeks to show how the insights of Goffman might contribute to an understanding of the nature and significance of these practices.

Key words: curriculum vitae, autobiographical practices, self, surveillance.

Introduction

In this brief article we develop the notion of 'auto/biographical practices' through the critical examination of an increasingly familiar element of professional life, especially academic employment, namely the curriculum vitae. We do this partly in the belief that the term 'auto/biographical practices' denotes a range of activities considerably wider than conventional understandings of the terms 'biography' or 'autobiography'. Such practices, we argue, are deployed whenever persons are called upon to tell a story about themselves or, in Goffman's familiar terminology, to engage in the 'presentation of self' (Goffman 1971). Actors are not called upon to 'tell a story' in the abstract; they do so before specific audiences and upon specific occasions. Such stories are judged, by presenters and audiences, as being adequate or otherwise for the particular circumstances which called for their presentation rather than with reference to any wider, more abstract or absolute notion of truthfulness or adequacy. It follows, therefore, that an examination of auto/biographical practices tells a disinterested observer as much about the circumstances under which such practices were deployed as about the individuals being described by such practices. This is certainly the case with the curriculum vitae.[1]

The present article should be seen primarily as a critical sociological reflection upon our own practices and, as a consequence of these explorations, a call for more systematic research in this area. Obviously we have both been involved in the production of CVs and the assessment of the CVs of others, either formally or informally. One major source for our explorations is the *Guidelines on the Presentation of a Curriculum Vitae* issued by the University of Manchester, the institution within which we are both employed. For obvious ethical reasons we have not directly quoted from CVs to which we routinely have access, although some colleagues have kindly let us see their CVs for the purpose of the present exercise. It will be obvious, also, that we are confining our discussion here to *academic* CVs; it remains to be seen whether different issues arise in other settings.

Contexts

Any set of autobiographical practices exists in a particular historical context, and the production and evaluation of CVs is no exception. We are here not so much concerned with the histories of CVs in general, although that would be an essential part of any more detailed analysis, but with a more recent and detailed focus on the CV as part of an overall system of institutional surveillance and rationalisation (Wernick 1991:154–180). The increasing formalisation of such procedures may be seen as reflecting the introduction of North American models of academic management and, more immediately, the increasing demands on the part of government and public bodies for greater accountability and rationalisation. The modern academic CV, therefore, may be seen as part of what Stephen Ball, writing with specific reference to schools, calls 'management as moral technology' (Ball 1990).

In the context of education, the CV and its scrutiny may be linked to another aspect of these wider processes of rationalisation and accountability, namely the process of appraisal and the appraisal interview. Typically, such interviews involve the scrutiny of CVs or some kind of equivalent document, although with a specific emphasis upon past achievements and future aspirations. Ball's account clearly has some applicability to university settings:

> The appraisal interview has elements both of the confessional and the psychoanalytical encounter, both of which rely upon the dynamics of self-revelation. The appraisees are encouraged to display their shortcomings, to seek out or to identify appropriate therapeutic procedures, and to judge themselves and award their own punishment (Ball 1990:161).

What is important here is the linking of the appraisal interview with notions of the self. But as well as this, the reader is encouraged to think beyond any particular managerial innovation to a wider consideration of the dialectics of power in modern institutions. Such linkages, we argue, should also be possible in the case of the CV such that it may be seen as revelatory not simply of the history of the individual attached to that document but also, through a process of critical reflection, of the historical context of its production.

It is worth emphasising the recent character of these administrative innovations. When one of us (Morgan) first applied for a teaching post at this institution in the 1960s, all the information that was required could be included in a brief application form and an accompanying letter. Now a letter, unaccompanied by a CV, would almost certainly fall at the first hurdle, unless the person concerned were already very well established. The significance of these changes, however, is not simply the simple fact of the arrival of the CV as an essential adjunct to academic careers. It is also the fact that the production and scrutiny of CVs form an increasingly central part of routine academic life such that it is now impossible to conceive of that life without CVs.

Theoretically, the context of this present article might seem to be unambiguously Foucauldian. However, while (as Ball demonstrates) there is plenty of scope for detailed analysis within such a perspective, our approach is more heterodox. From Goffman, we derive ideas of the presentation of self and impression management (Goffman 1971). Despite the fact that Goffman was concerned with face-to-face encounters, we found that there was plenty of scope for the application and development of these ideas in our own analysis. In this context we note the Goffmanesque connotations in the increasingly widely used term, '*performance* indicators', one major record of which being the CV. From a broadly ethnomethodological tradition, we derive notions of accounting procedures and indexicality. Here we focus upon the essentially interactional character of the production of CVs. CVs are produced and read in particular contexts and with reference to specific others and these processes of reading and writing cannot be understood outside these contexts. Put another way, the CV like any other administrative document, is never complete: it is always open-ended. Others are encouraged to read the gaps or to listen to the silences. For example, considerations of promotion to Senior Lecturer or beyond raises questions of the presentation and recognition of 'exceptionality'.

There are other influences at some distance from the mainstreams of sociological theorising. We share, with other articles in this issue, an interest in wider discussions of biography and autobiography in the broadest senses of these terms. For example, we make reference to a distinction made by Stenson (Stenson 1989) between thematic and rhapsodic narratives, a distinction derived from the analysis of social work interviews. Thematic narratives might be described as rational narratives: they are clearly orientated to a more or less articulated goal or goals; produced under a set of guidelines or constraints; they often require others or sanctions to 'keep to the point'; and their production has in some measure to be learned. Rhapsodic narratives, on the other hand, are less apparently rational in structure; less guided by obvious rules or sanctions; and generally less predictable. In rhapsodic narratives attention is drawn more to the present moment rather than to the end product.

Finally, we have been very much influenced by a climate of enquiry which seeks to take issues of experience seriously (Stanley 1992). Such approaches have clearly been influenced by feminist scholarship and the recognition of the gendered character of all social interaction; certainly, the production of CVs themselves has a gendered character. We see sociology as properly engaging with experience rather than separate from it; on the one hand requiring the constant critical scrutiny of experience, on the other seeing experiential accounts as valid and valuable sources of data.

Analysis

Some general observations may be made about the production of CVs and their subsequent scrutiny. Firstly, they are clearly and increasingly rule-governed. Guidelines exist in our institutions outlining how these are to be constructed and what is to be included or excluded. Graduate students may routinely expect to receive some guidance as to the construction of CVs as part of their normal socialisation into the profession. Often, more informal practices exist within departments whereby individuals are given advice on how to present themselves in the best light in the course of constructing their CVs.

The guidelines and formal and informal advice invite the writers (and, by implication, the readers) to steer a path between the public and the private, the relevant and the irrelevant, the general and the specific, and between quantity and quality. Such guidelines, in common with all such guidelines, can never be expected to cover everything and hence need to engage or draw upon the writer's own tacit and accumulated understanding of institutional expectations. Thus the sets of guidelines must be seen as part of a process of autobiographical construction rather than a simple blueprint or point of departure.

It should be noted, further, that the CV is not a straightforwardly linear account. While some temporal sequencing is required, perhaps in conformity with commonsensical constructions of individual life-courses as well as with administrative understandings of the cumulation of experience and expertise, the CV as a whole does not fully conform to conventional biographical structures. Certain periods of life (normally those prior to entry into a particular career track) are dramatically shortened and curtailed. Further, the writer is required to present temporal sequences under separate headings such as Research, Publication, Teaching and Administration. At the same time, the writer is not invited to provide a more lateral account for any one particular period, for example showing interplays between these headings for any given academic year. Thus, conventional understandings of a life course are fragmented according to administrative considerations and, similarly, certain periods of life are given markedly greater weight than others.

We have stressed that the production of the CV is rule-governed behaviour and that this feature has been growing in significance so that the rules have become increasingly explicit and formalised. It might be supposed that to the outsider one CV would look very much like another, differing only in length and 'density', i.e. the number of items in a given year. However, a somewhat paradoxical feature of 'CV work' is that it is also concerned with the presentation of a self, a process which might be described as the transformation of quantity into quality. Hence Wernick is correct to point out that the elaboration of the CV is not a simple function of the 'publish or perish' syndrome (Wernick 1991:169). Indeed, production of a CV takes place, as we have already noted, between two worlds. On the one hand there is the traditional academic world where quality is supposed to elude quantification and where the mysteries of a craft are embedded in sets of inter-personal understandings and invisible colleges. On the other hand, there are the increasing pressures to emphasise quantity, whether it be in terms of the number of publications, the size of research grants or the number of 'all expenses paid' international gatherings. The latter has not entirely replaced the former and out of this ambiguous co-existence rules and codes of practice arise in an attempt to codify the mysteries and the elusive quality.

It is here that we find Goffman's analysis of 'performance' to be particularly useful (Goffman 1959:17–76). Here, Goffman identifies eight elements in performance and, while not all of these elements may be equally applicable to the production and evaluation of CVs where self and others are not in direct verbal communication, these themes are of relevance to our analysis.

1. In the first place, Goffman notes a 'belief in the part one is playing'. Thus, while academics may frequently refer to activities such as the writing and presentation of CVs as a game or as a ritual, it is essential for an academic seeking promotion or tenure to present as a sincere performer. Jokey or satirical titles of published articles[2] are not favoured since these may indicate a degree of cynicism about the academic enterprise. It might also be noted here that a failure to conform to the rules of CV production through, say, the production of a skeletal or 'badly' organised document might also be read as indicating a lack of appropriate seriousness.

2. Another feature which Goffman highlights is that of 'front':

> the expressive equipment of a standard kind intentionally or unwittingly employed by the individual during his (*sic*) performance (p. 32).

'Front' involves selection and it is possible to see the formal guidelines on CV compilation as containing coded indications of those aspects of front to be employed. Thus Item C11 of the *Guidelines on the Presentation of a Curriculum Vitae* issued by the University of Manchester refers to 'academic invitations received for the presentation of external lectures and papers, contributions to conference proceedings etc'. Presenters of CVs are advised to 'draw particular attention to national or international occasions, and indicate which invitations were on an "all expenses paid" basis'. Here the concern is with the establishment of a reputation, an indication that the candidate's horizons extend beyond a particular university or department. The *Guidelines* seek to indicate which items might be of particular relevance in the establishment of a professional, national or international, reputation.

The establishment and the maintenance of a 'front' must be detailed without being trivial. Whilst it is important to provide a full and informative curriculum vitae, with no significant gaps or omissions, it is also important not to produce an over-lengthy c.v., full of trivial detail (Guidelines). The reader of these Guidelines must, of course, already have an understanding of what constitutes 'trivia' in order to make sense of these recommendations. What seems to be important is the production of a sustained front, produced in the recognition that unseen others will be scrutinising the document for gaps while being ever alert for the candidate who protests too much. A possible example may be found in the case of 'book reviews'. An academic at the beginning of a career may be allowed to include details of book reviews since this will indicate the beginnings of the establishment of a reputation. When more established, however, book reviews would normally be seen as a more routine part of academic life and such details, if provided, may come close to the classification of 'trivial'. Thus the front which has to be produced or sustained can be seen as varying over an academic life course. Now that book-reviews may in some cases be included among a set of 'performance indicators', it is possible that we may see their establishment as a regular feature of an academic CV with the consequent shift from quality to quantity.

3. Goffman next refers to 'dramatic realisation':

> the individual typically infuses his activity with signs which dramatically highlight and portray confirmatory facts that might otherwise remain unapparent or obscure (Goffman 1971:40).

This dramatic realisation presents particular problems when the communication is at a distance and the actor has no opportunity to modify a performance in relation to the responses of others. However, it is important that this document gives signs of something outside the routine and the ordinary, despite the fact that the format of the CV is becoming increasingly standardised. Academics have to exercise fine judgement here. Under the section on 'Administration', the University of Manchester Guidelines offer the following advice:

> It is important to stress major executive responsibilities within the Department as well as service on Faculty and University Committees, and any innovative activities. However, it is equally important to keep a sense of proportion and not to overstress duties of a routine nature which are undertaken by all members of staff from time to time (Guidelines).

Since there is often a very fine line between the routine and the exceptional when it comes to administration, the possibilities of dramatic heightening are considerable. However, similar fine judgements apply to all areas of the CV and it is here, among other places, that there may be elements for the more rhapsodical narrative to enter into a dominantly thematic account. Elsewhere, the opening may be provided for rhapsodic elements where the author is invited to provide expanded statements of research, teaching or administrative interests.

4. Goffman next explores the aspect of 'idealisation', the way in which and the extent to which a performer 'will tend to incorporate and exemplify the officially accredited values of the society'. In preparing the CV in the correct manner, an individual has shown a willingness to put time and effort into a particular mode of self presentation, one which in many ways exemplifies the

values of the wider academic community. Comments are sometimes passed on the extent to which a candidate has spent time and effort in the preparation of this document, the extent to which, for example, refereed articles have been carefully and correctly distinguished or the degree to which different categories of conferences attended have been indicated. Such comments, of course, parallel the comments which may be made about essays, dissertations, doctoral theses and articles submitted to learned journals, all pointing to measures of socialisation, anticipatory or otherwise, into an academic culture and into academic values. Thus the CV serves to celebrate and to reproduce the culture of the academic community not simply in the amount or quality of work indicated, but also in the demonstration that correct procedures have been followed and that due care has been taken.

5. Under the next heading, 'the maintenance of expressive control', Goffman shows that performances should not allow unwelcome intrusions from other areas of life into the front stage of the on-going presentation. In the case of the CV, one possible example of this was the advice offered to one of the authors about the entry headed 'marital status'. It was suggested that if she were to put down 'divorced' this might be read as 'making a political statement'. The same source of advice also suggested that references to temporary jobs such as shop assistant, clerk, etc., should be deleted as, here again, these might 'give the wrong impression'. Similarly, the other author considered and decided in favour of including National Service in the CV since the otherwise unexplained break of two years might be an occasion for comment.

There is, indeed, considerable uncertainty around the inclusion of such items. The section on 'Promotion to Senior Lecturer' in the Manchester University's *Manual for Heads of Department* includes the following advice:

> Information about nationality and marital status is optional: some candidates may be happy to give it, in the interests of providing a fuller personal picture, others may prefer to exclude it.

It is doubtful whether such advice, if passed on to the candidate in this form, would be all that re-assuring. Whichever, provided or witheld, the candidate has little direct understanding as to how such information or its absence will be read.

6. With Goffman's next heading, 'misrepresentation', we are approaching some very tricky areas relating to trust in academic life. He suggests that:

> many performers have ample capacity and motive to misrepresent the facts; only shame, guilt or fear prevent them doing so (Goffman 1971:65).

Such opportunities exist within academic life, especially as subject areas become more and more specialised and the acknowledged experts become more widely distributed geographically. Theoretically, it is possible to include fictitious articles, books, even journals, and certainly there are stories told of fraudulent claims which are only discovered after the guilty party has achieved relatively high or established status. It would be difficult to measure the amount of fraud in academic life, but here, as with clear cases of plagiarism, its existence is both possible and known.

Again there are fine lines to be drawn. While we have no evidence of anyone being advised to tell lies in the construction of a CV – indeed such a recommendation would strike at the heart of the nexus of trust that is central to academic and other professional cultures – there are occasions where some measure of circumspection might be exercised. In questioning each other on our respective CVs we invited ourselves to 'spot the whopper'. While we both failed in this task, further questioning did reveal some economies with the truth and one of the authors revealed being given the, doubtless, informal advice that 'what you need is some whopper training'. Thus if there be a measure of doubt as to whether a journal is a refereed publication or not, a candidate might be told to err on the side of the favourable. Alternatively, one of us included not only the number of post-graduates supervised but also listed those who had successfully completed; the other felt that this was overdoing the honesty. It was also wondered whether a 'warts and all' extract from a student evaluation of a course would be read as admirable honesty and a willingness to learn from mistakes or, more simply, as a sign of weakness. The point, however, is less whether such practices actually take place or, if so, their actual incidence. It is more that a reflection on the possibility of such practices highlights the significance of trust in academic culture and the way in which the CV might be one of a series of sites where trust is displayed or, sometimes, flouted.

7. Goffman's next heading, 'mystification', would seem to be of little relevance to the present analysis. While the production of the CV and its evaluation typically involves spatial and temporal distance, it does not as a rule involve social distance. This is because performer and audience generally share the same culture, are operating under the same set of rules, and are aware in some measure of the game-like features of the presentation in which they are jointly engaged. However, at the same time, the effective CV must be able to contain some awe-inspiring elements through the suggestion that the author inhabits scholarly worlds other than those of the assessors, worlds in which the author is honoured and recognised. As we have noted, academic life has not become completely routinised and there is enough left of the craft element within academic practice to enable some measure of mystery, and mystification, within the CV.

8. With Goffman's final heading, 'reality and contrivance', we are back to the heart of the business of self-presentation through the CV. Performers here have to give the impression of being a 'proper academic' or a 'proper scholar'. The task facing the academic in writing a CV is that of the presentation of an academic self, not simply in the 'real' achievements but in showing, through the display of these achievements, that one is a proper member of the academic community. Put another way, the 'reality' of these achievements to a large measure lies in their being read into the presentation of the text of the CV.

The academic, therefore, has to tread a narrow path between over-presentation of an academic self in the CV and failing to bring off an adequate performance in the preparation and presentation of this document. The ideal CV, one might suggest, would be close to the 'readerly' text whereby the readers and evaluators are persuaded to move relatively easily from the written word to the academic self and career thus displayed. Apparent absences, mis-representations or overstatements will impede this easy passage and may call into question the authenticity of this performance.

In this article we have concentrated on one section of Goffman's influential work in order to suggest its relevance for a kind of performance which might be thought to be some distance from Goffman's own interests. However, there are probably other features of this work which might be relevant, especially the sections dealing with 'team work'. Superficially, it might appear that the production of the CV is an individualistic act performed in a particularly individualistic context. However, not only is there scope for team work in the actual production of the CV, with the seeking or offering of advice on the part of senior colleagues or mentors, but there is also scope for teamwork in the subsequent evaluation. This is not simply a reflection of the fact that CVs are conventionally examined by committees who collectively pool their accumulated understandings of the meaning and production of CVs in the evaluation of a particular case, but also of the fact that performer and audience alike, although distant from each other, share a common culture and a common set of rules and expectations.

Concluding Remarks

We hope to have demonstrated that reflection upon the production and evaluation of CVs is not simply a matter of parochial interest but raises the possibility for the exploration of a whole range of auto/biographical practices. In focusing upon the CV as one of the kinds of documents that are routinely and increasingly called upon in the business of administering complex modern institutions, we are inevitably moving into areas that fall naturally into a Foucauldian framework, areas to do with surveillance, the exercise of the administrative gaze and the deployment of power in highly specific sites. We are not unsympathetic to such an approach. However, since issues of the self and its presentation are also involved, we feel that the work of Goffman is also highly relevant, especially when it comes to examining the more processual details.

We are aware that in discussing the interplays between the construction of a self and the production of a CV we have only been considering part of the process. We have left out of our consideration the production of references and their evaluation and the use of personal knowledge, both of which may be deployed in the evaluation of a particular CV. We would also need to say more about the appraisal interview. Clearly the importance of all these practices lies in their interaction rather than in their separate impacts. Ultimately, the evaluation of the CV cannot be conducted in isolation from these other practices. Put another way, it is important to consider the impact of CV production, and all these associated practices, upon the wider and changing academic culture.

These other features of the academic culture also involve biographical practices. This is clearly true in the case of the preparation and evaluation of references, but it also appears when the referees are themselves actually evaluated. Thus it is common for evaluators not only to read and to assess the references in relation to the candidate and to the CV on the table, but to also assess and to 'place' the referee. Thus, in the context of academic evaluation, we often have an interweaving of biographical practices.

As with all auto/biographical practices, it is important finally to speculate on what is left out of the account. It is here that we come upon what Wadel has called 'the hidden work of everyday life' (Wadel 1979). This emphasises the point that in any occupational setting, alongside the formally prescribed or routinely recognised practice, there also exist practices which are often vital to the maintenance of the organisation concerned and the cultures within it but which are not conventionally recognised. This might, for example, include a great deal of 'emotional labour'; the listening to complaints or concerns by peers or others; or the elaboration of forms of more or less pure sociability of the kind that contribute to the maintenance of some kind of corporate identity. It need hardly be said that such hidden work is often highly gendered and it is perhaps not coincidental that such work does not readily find a place in the construction of CVs.

The recognition of such 'hidden work' serves as a reminder of a growing contradiction within the production of CVs. On the one hand, the CV like any other such document is a one-sided accentuation of certain facets of self to be presented before particular and highly selective reference groups. It cannot be, and does not claim to be, a fully rounded construction of a 'whole person'. Nevertheless, academic work identities, in common with many other work and professional identities, do have strong linkages with the construction of personal identities and within this, as we have seen, the construction and constant revision and monitoring of CVs takes on ever increasing significance in terms of time and effort. Thus the self comes to be invested and constructed through a document that cannot ever be identified with any understanding of a 'real self' on the part of the individual actor concerned. Thus there is an increasing element of alienation in the production of CVs although the extent and nature of this alienation, especially in relation to the different experience of women and men within academic life, needs further investigation.

We hope that we have demonstrated the importance of seeing the presentation of self not only in terms of face-to-face encounters but also in the less direct encounters that are increasingly characteristic of large-scale organisations. We hope also to have shown, through developing the idea of autobiographical practices in the context of this particular extended case, that biography is not a specialised undertaking but something in which we are all engaged. It is through such practices that selves are presented and constructed. The CV, like many such documents in a modern organisation, is both a resource for administration and a topic for investigation and scrutiny in its own right (Scott 1990:36–7). We look forward to more detailed analyses of such practices.[3]

Notes

1. It is perhaps of interest that the recently published first volume of Muriel Spark's autobiography is called *Curriculum Vitae*.
2. For example, consider the title of the paper by Gilbert in the bibliography.
3. We note the publication, since our completing this article, of Andrew Metcalfe's examination of the curriculum vitae (Metcalfe 1992). This paper adopts a more thorough-going Foucaldian approach than the one adopted here; nevertheless, its almost simultaneous appearance is an impressive testimony to the growing centrality of the CV in various occupational cultures and the need for critical theoretical examination of this rise.

References

BALL, S.J. 1990. 'Management as moral technology: A Luddite analysis' in S.J. Ball (ed.) *Foucault and Education: Disciplines and Knowledge*. London: Routledge.
GILBERT, S.J. 1990. 'Scholarly image enhancement through a meaningless publication'. *Journal of Polymorphous Perversity* 7, 1:20–21.
GOFFMAN, E. 1959. *The Presentation of Self in Everyday Life*. New York: Doubleday Anchor Books.
METCALFE, ANDREW. 1992. 'The curriculum vitae: confessions of a wage-labourer. *Work, Employment and Society* Volume 6, 4: 619–641.
SCOTT, J. 1990. *A Matter of Record: Documentary Sources in Social Research*. Cambridge: Polity.
SPARK, M. 1992. *Curriculum Vitae*. London: Constable.
STANLEY, L. 1992. *The Auto/Biographical I: The Theory and Practice of Feminist Auto/Biography*. Manchester: Manchester University Press.
STENSON, K. 1989. *Social Work Discourses and the Social Work Interview*. Ph.D. Thesis, Brunel University.
WADEL, C. 1979. 'The hidden work of everyday life' in S. Wallman (ed.) *Social Anthropology of Work*. London: Academic Press.
WERNICK, A. 1991. *Promotional Culture: Advertising, Ideology and Symbolic Expression*. London: Sage.

Biographical notes: NOD MILLER lectures in the Faculty of Education at the University of Manchester. She has particular interests in experiential learning, feminist research and epistemology and media sociology. DAVID MORGAN is a Senior Lecturer in Sociology at the University of Manchester. His main interests are in the sociology of gender and masculinity, family studies and the uses of auto/biography in research and teaching.

Address: Department of Sociology, University of Manchester, Manchester M13 9PL.

STUDENTS' PERCEPTIONS OF REFERENCE LETTERS

Brian K. Payne, PhD
Sociology and Criminal Justice
Old Dominion University

Jonathan Appel, PhD
School of Criminal Justice and Social Sciences
Tiffin University

Donald H. Smith, PhD
Sociology and Criminal Justice
Old Dominion University

Kara Hoofnagle
Sociology and Criminal Justice
Old Dominion University

> This study examines students' perceptions of reference letters. Students (n=444) were asked to describe how they perceived reference letters. Four themes were uncovered. First, some students perceived reference letters as useful for employers. Second, some students perceived the letters as important for students seeking employment or admission to law or graduate school. Third, some students identified specific problems they had with the letters (e.g., distance education students). Finally, some students perceived the letters as entirely useless. Implications are provided.

The purpose of a college education for most students is very practical—they want to prepare themselves to get a job (Leppel, 2005; Payne and Sumter, 2005). While they may not know what job they want, which career is right for them, or how to go about getting the job, by the end of their four (or five or six) years of college, what they want out of their education is gainful employment. Employers base their hiring decisions on a number of factors including the applicant's work history, success in the educational arena, and potential for success in a specific job. Measuring the potential for success is not an easy task. Job references are just one tool that is used to determine such potential.

Most of what has been written about job references has focused on legal issues surrounding references for those providing references about certain individuals (Hirschfield, 2004; Picker, 1992; Rovella, 1995; Taylor, 1998), strategies employers can use to conduct effective reference checks (Andrica, 1998; Picker, 1992), and what employees can do to improve their likelihood of getting the job they want (Pechter, 1997).

The legally-oriented articles address the risks for liability and strategies to avoid being sued. One recent legal issue centers on the use of email to share reference letters. Questions center on whether such a process violates the Data Protection Act of 1998 (Taylor, 1998). Another legal issue related to job references entails the ability of states to pass laws encouraging employers to give accurate references (Rovella, 1998). More recently, attention has centered on ways colleges and universities are liable for references provided by instructors, professors, and other university employees (Hirschfield, 2004).

Research on effectively conducting job reference checks has centered on a number of different issues including the appropriate techniques to use, the types of questions to ask, and the relative weight to afford to references. One author considered the way employers should go about interviewing references, rather than relying solely on letters provided by the references (Andrica, 1998). In particular, among others, Andrica advises that the following questions should be asked of potential employees' references:

- What are this person's strengths?
- Can you cite an example illustrating these strengths?
- What accomplishments or successes did this individual demonstrate in your organization?
- If you were to offer any advice or constructive criticism, what would you tell this person? (p. 132)

The responses to these questions are believed to be quite helpful in deciding who to hire.

Research on the usefulness of job references has centered on strategies employees can use to improve their likelihood of being hired (Pechter, 1997). According to Pechter, networking will help individuals better prepare themselves for subsequent employment. One employment consultant advises that job seekers improve their chances of employment by:

- Asking past employers what kind of reference they will provide
- Confirming references in advance
- Calling references before a potential employer calls them
- Not overburdening references by listing them for jobs in which the applicant has no interest
- Offering to serve as a reference for one's references (*OfficeSolutions*, 2003).

While research has examined aspects of job references related to employers' liability, strategies to use them appropriately, and ways employees can use references effectively, very little research has considered how students perceive job references. Such research is useful because it will help determine if students' perceptions of job references will help or hinder their abilities to gain employment, which as noted earlier, is the ultimate goal of a college education.

Methods

Students from Old Dominion University (n = 444) completed a voluntary survey conducted by the members of the Department of Sociology and Criminal Justice Department at Old Dominion University. The majority of the students that completed the survey were criminal justice, sociology or human service majors. Respondents were undergraduate students, ranging in age from seventeen to fifty-seven. The survey consisted of several close-ended questions relating to different social problems. At the end of the survey, students were asked to describe their perceptions of reference letters. Five close-ended questions were included as well as one open-ended question: What are your thoughts about reference letters? For the purpose of this analysis, we content analyzed their responses to this question.

Findings

Overall, reference letters were found to be an overall positive element. Responses such as "They are great," "They are good," "They are helpful," and "I feel that reference letters are a useful tool," and "They are needed" are some of the comments that were expressed to show the positive views that students have about reference letters. Beyond these general findings, our content analysis revealed four themes characterizing students' perceptions about reference letters. These themes included (1) useful for employers, (2) useful for students, (3) problematic for some students, and (4) useless for all.

Useful for Employers

Perhaps the most common theme was that students tended to view reference letters as tools useful for employers. For example, one student said, "They are a good way for employers to see how others in society view your character." Making virtually the same comment, another student said, "They are good to use as a character witness of the applicant." Below are a few other examples of comments illustrating the useful for employers themes:

- I think they are important part when an employer decides whether to *hire* you or not.
- I think they are good, gives you a little more about *what the person is like*.
- They just provide employers *with info about the person* from another person's perspective.
- Reference letters give an employer or a college a better idea of the *personality and work ethic of the applicant*.
- They are a good way for employers to see how *others in society view your character*.
- They are important because they enable the future employer to *get to know the person through someone else's eyes*.

Some students relied on their own experience as managers or supervisors in developing their assessments about reference letters. Three students with past experience hiring students made the following comments:

- I think reference letters are a good way for employers to get to know their applicants. It's the one true way to get to know who they are besides calling the references. I think reference letters are a positive thing.
- I've hired people before with both good and bad references and the ones with good references are always the good employees. The ones with bad references are always the ones that quit right away or get fired. I think they are a great tool.

- I think it gives an employer an idea of what type of person you truly are. A 4.00 GPA doesn't tell as much as a written statement from someone who has dealt with you on a professional level.

Useful for Students

A second theme that arose is that many students tended to see reference letters rather introspectively, or as tools that will help them. The areas in which reference letters were cited as being most useful included gaining admission to college and attaining employment. Said one student of reference letters, "They help students get in to better college." Another student said, "They are important for getting into graduate school."

Indeed, comments such as "They help you get the job" and "I think they are very important when it comes to getting a particular job" were common examples of the usefulness to students theme.

A few other students expanded on why they thought the letters helped them get jobs. Here are a few of examples of these more detailed assessments of job references:

- These letters are very important to me and they must reflect the qualities that I have. I also like to know how people think about me and it may give me a chance to enhance some of my qualities
- Reference letters are very crucial when it comes to getting a specific job that meets all the criteria and expectations that you personally want and need.
- They are important because it gives an idea on *what you are like* and *how well and how hard you work from a respected member of society.*

Problematic for Some Students

Another theme that arose is that some students felt that reference letters were problematic in specific contexts or situations. Those situations which were problematic were case in which students (1) have a high employment potential but no faculty contacts and (2) are enrolled in distance education programs.

With regard to the first group of students who would be good employees but have no faculty references, some students implied that the job references do not adequately address the true nature of the student as a future employee. Said one student:

Reference letters are too highly regarded. A person could be more capable for the job and have no references and some nobody has been could have references and get the job.

Another student made a similar comment stating, "Maybe you have no references; it doesn't make you less valuable than another." Other comments made by students which seemed to indicate that they have the same or similar perceptions of reference letters include the following statements by four different students:

- They only show positive aspects when sometimes the negative needs to be shown.
- Adds positive things to resume but not overly important, good to have though.
- Reference letters don't always portray

or reflect who the person really is.
- I think reference letters are somewhat skewed because they are usually written by a person who likes the one who the letter is from, and the writer will probably tell the good qualities of a person.

With regard to distance education and reference letters, unique to Old Dominion University is their TELETECHNET program. This program is a distance-learning program. Although effective, in terms of reference letters, this program proposes challenges in terms of obtaining reference letters. The following is some of what TELETECHNET students had to say about reference letters:

- Being in a TELETECHNET class, I feel that a reference letter won't do too much good.
- I am a TTN student so it would be harder for me to request reference letters from someone who doesn't know me and has never met me. Plus you may never contact the professor but only their graduate assistant. I wonder what the letter would say. All they have for their opinion of me is my grade.

Although some TTN students feel that they may have a hard time acquiring a reference letter from an employer, their overall comments were similar to on campus students in that some found them beneficial, and others found them to be biased and not important.

Useless for All Students

Additionally, another theme that emerged from the surveys was that many students find reference letters biased, not important, and a waste of time. Said one student, "They are of little value." Another student made a similar comment stating, "I question their effectiveness." Another student said that reference letters are "inaccurate and biased." Other comments displaying this theme of the uselessness of references letters include the following comments from two different students:

- They are a waste of time and effort. You only get them from people that you know will give you a good one. You always come across as good.
- I'm not sure if they serve a great deal of purpose. No one is going to ask a professor to write a reference letter unless they know it will be positive.

In summary, the majority of students responded that they think that reference letters are good, helpful, necessary, and important. Similarly, the majority also expressed that reference letters focus on the positive aspects of the individual. Some viewed this particular aspect as an okay thing, while others found it the factor that makes reference letters biased, and negative. Overall, student's responses indicate a type of guarded optimism towards reference letters, those that write them, and those that read them.

Discussion

The results of our study are interesting and suggest that students have mixed perceptions about the usefulness of reference letters. Some students held reference letters in high regard for either employers or students, others noted problems they had with reference letters, and other described reference letters as useless. Based on our findings, a number of recommendations can be made so that professors can better utilize reference letters for students.

First, students should be made aware of the usefulness of job references. Related to the importance of references is the whole process of networking. In many cases, employees are given opportunities to interview for jobs because of who they know, not what they know. Incoming freshmen need to be made aware of the importance of getting to know their professors so that they are able to ask them for references. Learning the trade of networking with their professors will help students network with others.

Second, measures should be taken so that students who encounter problems with reference letters can be better represented. Consider students enrolled in distance education programs. Strategies may need to be developed for distance education students so that they aware of how to seek out job references. Likewise, strategies may need to be developed for faculty members so that they recognize the importance of serving distance education students the same way as traditional students.

Third, students should be encouraged to seek reference letters even before they graduate. If professors are willing to provide them, students should request a general reference letter for their files, whether they are seeking a job, graduate school, or law school. Such letters fulfill two instrumental purposes for students. First, the letters provide students with the faculty member's insight into their potential for success. Second, and more importantly, if students wait too long (even a semester or two), the faculty member may very well forget about the specifics of the student's strengths and have problems writing a solid letter. If a general file is available, the student can either use that letter later in time when needed or supply the faculty member with a copy years down the road in those cases when an updated letter is needed.

Finally, to make the reference letter process more useful for students, professors should empower students to assist in the letter writing process. Some professors have asked students to write their own letters only to have the professor sign the letter. While this may work for advanced graduate students, undergraduates will likely be too timid to provide the letter. Moral and ethical issues also arise with this process. Instead, when we suggest empowering students in the process, what we mean is that students should be prepared to ask for reference letters appropriately and provide enough information so that accurate letters can be developed.

In particular, faculty should consider asking students for the following products before doing their reference letters:

- Student's updated resume. If the professor knows what the student has done and plans to do, he or she can speak about those activities with confidence.
- Pre-addressed postage-paid envelope from the student. This is not about the budget crisis. It's about making it easier to get the letter in the mail, especially

if the professor is working from home. Also, the professor can make sure it is being sent to the right place.
- Unofficial transcript. The professor wants to know what courses students have had, their minor, and their grades.
- List of classes the student has had with the professor. It is impossible for professors to remember all students. They need to be reminded about specifics.
- Reminder about the papers the student wrote in the professor's class. If the professor can say something about the student's writing ability, it will certainly help.
- Copy of the student's senior project if it is available. Professors can't make things up about the student's writing abilities. They need to be reminded with concrete examples.
- List of accomplishments students are most proud of. Professors need to know what students are all about. This will help professors make the letters more "personal" and relevant for employers.

Students should also be encouraged to only ask faculty members who they know will give them a good reference. There is nothing worse than getting a lukewarm reference. Students are probably better off with no reference than they are with a generic one. They should be told ahead of time by their advisors or instructors that refusals to provide reference letters should be seen as a favor professors are doing students.

In the end, it is important that students realize that references can be the "icing on the cake" that gets them their job or into graduate school or law school. It's not just the letter that is important; rather, it is the student's ability to network with professors so that he or she is prepared to network in their future careers. While a good reference may not in and of itself get someone their favored job or admission to graduate school, a bad reference will most assuredly seal their fate.

References

Andrica, Diane C. (1998). "Interviewing Job Candidates' References." *Nursing Economics.* 16(3): 132.

Green, Raymond, McCord, Mallory, and Taiwa Westbrooks. (2005). "Student Awareness of Educational Requirements for Desired Careers and the Utility of a Careers in Psychology Course." *College Student Journal, 39,* 218-222.

Hirschfield, Stephen J. (2004). "The Dangers of Employment References." Chronicle of Higher Education, June 25, p. B11.

Leppel, Karen. (2005). "College Persistence and Student Attitudes Toward Financial Success." *College Student Journal, 39,* 223-241.

OfficeSolutions. "Manage Your References for Job Seeking." 20(4):10-11.

Payne, Brian K. and Melvina Sumter. (2005). "Student's Perceptions of Career Fairs." *College Student Journal, 39,* 269-276.

Pechter, Kerry. (1997). "Job Hunting?" *Electronic Engineering Times.* April 7, 118.

Picker, Lauren. (1992). "Job References: To Give or Not to Give." *Working Woman, 17,* 21.

Rovella, David. (1995). "Laws May Ease the Risky Business of Job References." *National Law Journal, 18,* B1.

Taylor, Chris. (2001). "Email References." *People Management, 7,* 13.

*Reflections on the Academic Job Search in Sociology**

SHELIA R. COTTEN, JAMMIE PRICE, SHIRLEY KEETON,
RUSSELL P.D. BURTON, AND JANICE E. CLIFFORD WITTEKIND

Although a variety of resources exist to aid social scientists, in general, in their academic job pursuits, almost no resources detail this process specifically for sociologists. This research aims to provide a review of the literature on the academic job market and to provide a sociological analysis of the job searching process in sociology. We report and analyze the experiences of five sociologists as they moved from graduate school, post-doctoral fellowships, an applied research position, and a non-tenure track teaching position to assistant professor positions. We detail the importance of impression management and self-presentation throughout this process (by job candidates, departments, and administrators), the resources provided by the American Sociological Association that exist to aid job candidates, and we suggest implications of the current organization of the sociology job market.

Introduction

Job searching can be a very trying and exciting process for sociologists. However, few sociologists have examined what actually occurs during this process and if patterns exist that might contribute to our sociological understanding of the sociology job search process. Given this, the goal of this research is to (1) review the limited available research in this area and (2) further illuminate the sociology job searching process based upon an analysis of information provided by five sociologists who completed job searches during the 1998-1999 academic year.

A variety of materials exist to guide academic job seekers, in general. There are assorted books (Boufis and Olsen 1997; Kronenfeld and Whicker 1997; Heiberger and Vick 1992; Sawyer, Prichard, and Hostetler 1992; Zanna and Darley 1987) and a few articles (Hotaling 1994) offering job-searching suggestions that cut across social science disciplines. For example, Boufis and Olsen's

---

* Throughout this article we use the pronoun "she" to preserve the anonymity of our co-authors.

The authors thank Barbara Risman, Beth Rushing, and Jennie Jacobs Kronenfeld for advice and suggestions regarding this manuscript.

Address all correspondence to: Shelia R. Cotten, Department of Sociology and Anthropology, UMBC, 1000 Hilltop Circle, Baltimore, MD 21250; E-mail: cotten@umbc.edu.

(1997) and Sawyer et al.'s (1992) edited books include advice from the experiences of academic job candidates in such fields as English, Psychology, Math, Women's Studies, Medicine, Physics, Archaeology, History, Philosophy, Sociology, Art, Communications, Engineering, and American Studies. These contributors offer advice on applying for jobs, interviewing at conferences and campus visits, job offers, identity politics during the job search, underemployment, and alternative careers. Similarly, Zanna and Darley (1987) and Kronenfeld and Whicker (1997) extend interdisciplinary advice on how to find academic jobs, what to expect from campus visits, preparing for campus visits, and negotiating job offers. Hotaling (1994) advocates that graduate students, graduate faculty members, and administrators take strategic steps to cultivate non-academic job options for Ph.D. graduates.

In addition, there are several resources that target individual disciplines. To illustrate, Rheingold (1994) and Brems, Lampman, and Johnson (1995) target psychologists, Sheehan and Haselhorst (1999) social psychologists, and Stith (1990) family studies. And, there are numerous publications on the academic labor market for economists (Shively et al. 1999; Siegfried and Stock 1999; Kahn 1993; Carson and Navarro 1988).

While these publications certainly help, sociologists seeking academic jobs would best be served by publications that comprehensively detail the sociology academic labor market. Publications detailing the demand for academic sociologists, the supply of Ph.D. sociologists, and doctoral degree outcomes are becoming more available to guide job searches in sociology. The American Sociological Association (ASA), regional sociological associations, and the *Chronicle of Higher Education* publish much of this information (Broughton 2000; Spalter-Roth and Lee 2000; Spalter-Roth, Thomas, and Levine 2000; Magner 1997; *Chronicle of Higher Education* 1992, 1994).

Other sociologists have offered advice on graduate school training (Ault 1996), surviving conferences (Cohen 1997), dealing with rejection letters (Shaw 2000), and managing a career in sociology with ethical and moral fortitude (Marx 1997). The effects of racial and gender patterns on employment within sociology have also been examined (Misra, Kennelly, and Karides 1999). In addition, several biographical and autobiographical works have focused on the experiences of specific groups of sociologists, including women (Goetting and Fenstermaker 1995; Deegan 1991; Orlans and Wallace 1975) and working-class academics (Grimes and Morris 1997). Also available are biographies on the careers of successful sociologists, most published posthumously, in many of sociology's top regional, national, and international journals.

None of the above publications describe what actually occurs during the job search process for sociologists. Do patterns exist that might contribute to our sociological understanding of the sociology job search process? Building on the literature previously mentioned, we now attempt to further illuminate the sociology job searching process based upon information provided by five sociologists who completed job searches during the 1998-1999 academic year. While discussing our experiences, we highlight the sociological importance of identity work, self-presentation, and interactional patterns in the job search process. We begin by detailing the backgrounds of the five sociologists involved in this project.

Backgrounds

The five faculty members we describe represent a diverse background of sociologists moving into assistant professor positions. During the 1998-1999 academic year, two of us held post-doctoral fellowship positions, one held an applied research position in industry for four years, one was ABD, and one was in the first year of a non-tenure track faculty position at a teaching school. Our specialty areas overlap in terms of a focus on health and illness (n = 3), social psychology (n = 5), family (n = 2), and research methods (n = 3).

Four of us had previously published research, some more recent than others and some in higher tier journals. The number of our journal publications ranged from two to seven (average = 4.5), one person had published a book, and two people had received grants. Beyond having publications, we each had well-defined short- and long-term research agendas and were able to clearly articulate these agendas. Even at strictly teaching institutions, our research agendas were instrumental in getting us interviews.

We all received our Ph.D.'s from large state research universities. And, each entered the mid-tier academic labor market. Our number of campus interviews ranged from 1-6 and totaled 20, with three of us declining additional invitations for campus interviews after we accepted offers from other schools. Four of us obtained tenure track positions: two at large regional research universities, one at a mid-size public university, and one at a small public university. One person obtained a visiting assistant professor position at a predominately teaching school. Three of us obtained positions requiring us to teach more general topics than our substantive areas.

Our specialty areas strongly influenced our job search experiences. They determined the jobs for which we could reasonably apply. Criminology, health, inequality, and research methods have been the most advertised positions in recent years. There are few advertised jobs in social psychology and family, and even fewer jobs in more specialized areas of interest. For example, one of our members' second major area is in social networks while another's is in masculinity and sexuality, each of which is even less in demand than social psychology. Given that we rarely encountered positions that encompassed both of our major substantive areas, we often had to consider how well we could fit our interests into other advertised areas.

Identity Construction and Presentation of Self

One of the things that we learned about sociology through this process is that identity theory, self-presentation, and impression management come alive during the job search process. These processes are inherent within any discipline, but they are highly applicable to the discipline of sociology given the focus on these issues by Goffman, Mead, Stryker, and others. As sociologists, we were all aware of situational demands and of the many pressures to present a particular self, depending upon the type of position, the type of school, and the individuals with whom we were interacting.

Even before going on the market, we had to decide what our identity was as a sociologist. Were we generalists, hoping to obtain positions at teaching schools?

Did we more closely identify with the specialized teaching interests and high research expectations of research university positions? Did we see ourselves fitting into other types of settings, such as schools of public health? And, how would we perceive ourselves if we did not obtain a position in academia? Would we find non-academic positions professionally and personally fulfilling? Would we still be sociologists? What does being a "practicing sociologist" mean? These were important questions that impacted how we proceeded with our job search process and how we evaluated its outcomes.

As the job search progressed some of us went through the process of evaluating who we were, in a professional sense, based on the job description. First, we asked ourselves if we could identify ourselves as the type of sociologist in any given advertisement. Then we asked ourselves questions such as: "How do I want others to see me?" and "What initial impression do I want to make?" We used the answers to these questions to support our identity and structure our interactions, whether it be forming our cover letter and vita by incorporating specific words and phrases, or managing our actions when meeting faculty face-to-face.

For example, for a family sociologist position, we would be sure to mention our training and research experiences in this area. For generalist positions, we would highlight our training in research methods, statistics, theory, and introductory sociology. Most of us marketed ourselves as general sociologists for some jobs, specialists for others, and both for still other jobs.

Applicant Presentations of Self

Presentation of self for job applicants occurs at a variety of times during the sociology job search process. It occurs as candidates develop vitas and cover letters, during informal meetings with potential employers, and during more formal interviews (whether they are phone or face-to-face interviews). In order to make the best presentation of self, candidates must think carefully about the information they are giving off and the impressions they wish to make. In order to prepare appropriately, it is necessary to learn as much as possible about the departments with job openings.

For the five of us, this process became even more intensified once actual interviews were scheduled. We worked hard to learn as much as possible about the departments we would be visiting in order to make the best presentation of self. We utilized the Internet to find departmental and program brochures, university catalogs, course schedules, faculty publications, sex and race composition of department faculty, number of students, school history, and community resources. The ASA *Guide to Graduate Schools and Departments* was also helpful in gathering this information. A few departments mailed us information on the faculty, university, and community. Several of us were surprised at the lack of initiative on the part of departments to send us information on their schools. We found that we had to be very aggressive in requesting this information.

We also gathered information through our networks of friends and mentors who knew about the departments and/or the communities. Information on things like job satisfaction, turnover problems, tenure denials, and department problems, politics, culture, expectations, and collegiality are gained more easily through these informal networks than through the actual interview process. Professional networks can be particularly helpful in determining whether the department focuses on particular sociological issues, such as inequality or criminology, in its teaching, research, and service.

We used this information to (1) develop questions to ask on our campus visit, (2) inform our conversations during the campus visit, and (3) assess the department's activity level. In order to manage an effective self-presentation during campus interviews, our experiences indicate that, at a minimum, candidates should know about the programs within the department, faculty names and areas of research and teaching, and how the candidate could contribute to the department.

During campus interviews, we found that most schools scheduled one main interview day during which we were escorted from one meeting to another. Most of us asked the school to send our schedules to us before we left for the interviews. This allowed us to prepare questions for each individual whom we were scheduled to meet during the day. All of us also met people who were not on our official schedule. Hence, knowing something specific about everyone in the department, not just those on the itinerary, paid off. The more prepared you are, the more professional the self-presentation. For many of us this meant taking information about the faculty and department (including vitas, publications, degree requirements, etc.) with us to interviews. We studied the material during our travel time and while in our hotel rooms before and after meetings with departmental representatives.

All the information we gathered contributed to our ability to present a self that matched each department's needs and wants. In analyzing our experiences, we realize that our self-presentation ability varied as a function of several things. These included whether we were interviewing with (1) teaching or research schools, (2) traditional departments that offer only sociology majors, (3) mixed departments that offer other majors such as social work or anthropology, or (4) interdisciplinary departments that offer majors such as criminal justice, gerontology, or public health.

Managing our self-presentation was particularly difficult when we interviewed in mixed departments. In some instances, we met with faculty in the other disciplines but were aware that they would not be voting on our candidacy. While we wanted to make a good impression with each person we met, it often seemed that our interactions with these faculty members were cursory at best and they often showed little interest in our candidacy.

We were also very cognizant of our self-presentation in traditional sociology departments versus interdisciplinary programs, such as schools of public health. Some of these distinctions also correspond to differences between teaching and research universities. A majority of the institutions with which we interviewed were considered research institutions. When interviewing with a teaching school, we asked more questions about teaching (i.e., skill level of students, opportunities for implementing new courses, size of classes, on-line teaching, teaching load, number of preparations per semester, number of majors, teaching evaluation policies). When at a research school, we asked more questions about research resources and expectations (i.e., availability of research assistants, data analysis program availability, publication and grant expectations, faculty collaboration, travel money, university support services).

Two of the things that sociology job candidates are often required to do during interviews are to give a research presentation (also known as a job talk) and teach a class. Each of these allows the faculty to evaluate the candidate's self-presentation, communication skills, and ability to interact with faculty and possibly students. We discuss our self-presentation in each of these in more detail.

Job Talks. Preparation and practice were keys to presenting an image of self that corresponded to what departments desired. Given that the job talk is one of the most important determinants of whether a job offer is made (Darley and Zanna 1987), we took this part of the interview process very seriously and attempted to make the best presentation possible. From our experience, we offer the following suggestions concerning the job talk phase of a campus interview.

First, the job talk should be tailored to fit the department and the specialty areas in which they seek to hire. A job talk for a traditional sociology department should differ from one for a multidisciplinary department, and from one for a research university. For example, while strong methods and statistics are required for any job talk, candidates should particularly emphasize these for presentations in research universities.

A talk for a traditional sociology department should include a strong theoretical foundation. Many of the traditional departments we interviewed with asked specific questions about our theoretical grounding or perspective(s). One co-author did not detail the theoretical foundation in a job talk. Later, the candidate found out, from inside sources, that this was a key reason why the department offered the job to someone else. Needless to say, our co-author incorporated a discussion on theoretical foundations in all subsequent presentations to traditional departments.

Compared to more traditional sociology departments and teaching schools, we gave different job talks and presented somewhat different research agendas when interviewing at interdisciplinary departments or schools. For example, at schools of public health, our experiences suggest presenting very specific information on grant writing experience, current grant proposals in progress, and future grant writing goals, while downplaying theory and teaching unless specifically asked about them.

Second, regardless of the type of institution, a job talk on completed research is preferable to one on research in progress. Third, candidates should organize their talks into beginning, middle and end sections. The beginning section should introduce the topic. The middle section should discuss the heart of the project, usually the results if applicable. The end of the talk should relate the project to the bigger picture. For each of us, our goal was to *tell the story* during the presentation, not to just give numbers without meaning.

Teaching a Class. A few departments asked us to teach a class during our campus visit (n = 8), all but two of which picked the course for us to teach. In the majority of cases, we were able to choose the topics that we would teach. In all cases, the department faculty observed our teaching and interaction with students. We want to emphasize here, though, that most schools did not request that we teach a class. Some faculty members even alluded to the fact that as long as we could publish and weren't horrible in the classroom we would be fine. Schools of public health were even more extreme in their views. Teaching held minimal, if any, importance. Obtaining external funding was of utmost importance, followed closely by publishing.

Even if teaching skills were not ascertained through teaching a class, most sociology departments were still keenly interested in what classes we would like to and could teach. Obviously, it was important for us to discern what teaching areas they needed. We did so by relying on information from the job advertisement, from our mentors and networks, and from conversations with faculty members during the campus visit. This knowledge allowed us to present an identity consistent with what the department desired in terms of teaching abilities.

The small number of schools that requested that we teach a class illustrates the importance placed on teaching. Many of the institutions did not request a class presentation, thereby demonstrating a different set of priorities: research, publishing, and external funding come first. Our perceptions would most likely have been different if we primarily interviewed at "teaching schools." It seems paradoxical to us to be in an institution of higher education and yet place such little importance on teaching skills.

In summary, our presentations of self changed depending upon the type of institution and department with whom we were interacting. As sociologists, we wonder about the effects of attempting to manage multiple identities while remaining true to our authentic self (Brewer 1991; Gecas 1982). How does it affect who we are as sociologists and as individuals to continually change the images we wish to present within one specific identity—that of sociologist. At times, this management of identities resulted in confusion, as when we had to follow up with those we interviewed with earlier. In these settings, we had to be very cognizant of the previous images we presented. Although we saw our specialization areas as complementing each other, interviewers were more concerned that we fit their position description. Thus, noting how our research and teaching related to the areas desired by each school was key to decreasing the stress associated with this process.

The more interviews we had, the more skilled we became at managing multiple identities. But, sometimes our self-presentations were not as successful as we had hoped. For instance, one of our co-authors interviewed for a medical sociologist position and later found out that she wasn't offered the job because the department members thought she was a social psychologist trying to pass herself off as a medical sociologist. She later obtained a position as a medical sociologist, after learning to better manage her self-presentation. Although she had specializations in both areas, she downplayed social psychology and emphasized medical sociology in future interviews. Given that social psychology is more theoretical in many ways than medical sociology, she thought that interviewers would see the combination of specialty areas as a strength rather than as a deficit. She also reported feeling annoyed that departments did not understand how her research incorporated both these areas. Given that problems with impression management affect how we perceive ourselves and how others perceive us, preparing for campus interviews extensively and mindfully is a necessity.

Departmental Presentations of Self

Departments had a variety of opportunities within which to present themselves. These ranged from the person who made initial contact with us to request more information or schedule an interview, to the material that was sent to us about the department and institution, to the people we met and interviewed with during our campus visits. In general, department representatives at each phase of contact were friendly and attempted to put us at ease during our campus visits. A few exceptions to this will be detailed later in this section.

We met many faculty members during our campus interviews. Most of us were initially met by either a search committee member or the chair of the search committee. Although we were not able to meet with each person individually, a majority of department members participated in our campus visit by attending our presentations, eating a meal with us, or interviewing us one-on-one.

We did not sense that faculty members were striving to present a particular image of the department by dressing in a particular way. Faculty dressed in a wide range of styles, from denim jeans to ties and suits, but most dressed in a business casual fashion. Department chairs usually dressed more formally, in dresses, suits and ties, and none wore denim, unless it was at an evening social organized for us. Nearly all faculty were initially introduced with titles, including first and last name, and we were usually introduced similarly. At many schools, after initial introductions faculty referred to each other, and to us, by first name. And, with the exception of three schools, most faculty were punctual in meeting with us during scheduled times.

Regarding unity, most of the departments we interviewed with displayed harmony. Faculty interaction with other faculty was very friendly. We often observed faculty talking with each other in their offices. At most schools, large groups of faculty attended our job and teach talks as well as our group meals and group social activities. It became clear from the discussions during these interactions that a core group of faculty at several institutions ate lunch together daily and interacted regularly outside of work.

We observed disharmony in six departments, three of which were multidisciplinary departments. Disharmony was usually evidenced by discussions of infighting between department faculty. In a few of the multidisciplinary departments, sociology faculty often spoke poorly of their anthropology and social work colleagues. Likewise, as stated earlier, the anthropologists and social workers in these departments showed little interest in our candidacy, with notable absences from our campus interview activities. However, in the other multidisciplinary departments, the faculty of all disciplines participated equally in our interviews, attending group meals and conducting individual interviews with us. Faculty relations in these departments appeared harmonious regardless of discipline.

Despite their seemingly professional first impressions, we learned to expect the unexpected from faculty members. During the campus interviews, two faculty members asked questions that were not directly relevant to the job talk or specialty areas. For example, one faculty member asked one of our co-authors who did not specialize in inequality to detail the current state of the literature on race, class, and gender, and how she felt her work contributed to this body of literature. Needless to say, this question and the department member asking it reflected a less than polished department presentation of self. Our co-author later learned, year after year, this faculty member asks each job candidate this same question regardless of the candidate's specialty areas.

As with our own, departmental presentations of self also differed based upon the type of institution. In general, our experiences indicate that department members at teaching schools ask more questions about teaching, whereas department members at research schools ask more questions about research. We expected teaching and research schools to differ in their presentation of self along these lines. But, the presentation of self among some departments of high prestige shocked us.

For example, one of our co-authors perceived the faculty at a prestigious institution as elitist. The faculty acted as though they did not have time to spend with the candidate. The candidate was even left alone at lunch. A few faculty members made no attempt to meet the candidate during the campus interview. Our co-author reported feeling very disillusioned about the whole interview process at this institution. Given the state of the labor market in sociology at that time and the institution's prestige, it is probably no surprise that the faculty acted as they did. In the hierarchy of the academic marketplace, the faculty held the power position and the candidate, attempting to move up, the subordinate position. While sociologists are very good at studying inequality, they also appear to know how to practice it effectively.

Race and Gender. In retrospect, we recognized several dimensions of departments' presentations of self during our campus interviews that reflect larger issues of race and gender within sociology and within society. All five of our cohort members are white, and four of us are female. From job advertisements and our professional networks, we learned that many departments sought minority candidates. Departments made this clear in their job advertisements by encouraging minorities to apply. Others subtly conveyed their plan by, for example, noting their equal opportunity employer status in their job advertisements. Still others hid their intentions to hire a minority person behind their request for candidates specializing in the sociology of race.

Departments seek minority candidates for many reasons. Sociology, as a discipline, values equality and strives to represent gender and race equitably among department faculty. Most of the departments we interviewed with included few faculty of color, which, we imagined, made hiring a person of color a priority. Additionally, university administrations often provide special economic incentives that allow departments to make a higher salary offer to a person of color, and even gain a second position for the program if they hire a person of color. Aguirre (2000) provides further details on affirmative action issues.

Faculty and departments presented gender in several ways. At a few schools (n = 5) we observed that women faculty, as a group, spent more time getting to know us, and shared more information about themselves, the program, and the department. In short, these women faculty treated us more as insiders. In explaining these gender trends, it appears that a gender and age interaction occurs at times. One of our co-authors suggests that women were often the most junior, and youngest, faculty in the department, which may have induced more empathy among them for job candidates who are also junior and tend to be relatively young. Our observations may also mirror the social fact that women in our society provide a majority of care and emotion work (Weiss 1990; Shields 1987; Hochschild 1979). While these interpretations of the gender situation are reflective of several of the co-authors' experiences, we note that our small sample may not necessarily be representative of the interviewing experience, in general, at all institutions.

With regard to the race and gender distribution among department and university administrative personnel, we found, for the most part, virtually no representation of people of color and an overrepresentation of men. Overall, the departments included an average of 60 percent men, 40 percent women and less than 5 percent people of color. The search committees tended to be about 50 percent men, with about 50 percent chaired by men. None of the search chairs were people of color. Most of the department chairs (about 75%) were men, and only two of them were persons of color. For the most part, we did not notice a difference in presentation of self or institution among men and women chairs. Most tended to act polite and personable, regardless of gender, but a bit more distanced and formal than the department faculty.

Providing Department Information. In addition to issues of gender and race, we picked up subtle cues and conflicting signals about the departments both during and after our interviews. One faculty member warned about the heavy workload of the position, implying that the candidate should think twice about accepting an offer for the position. Another briefly mentioned the heavy service expectations in the department, but never disclosed the magnitude or manifestations of this service requirement. At several schools (n = 6) faculty members warned us about unclear tenure expectations, offering alternative tenure guidelines for us to follow instead. Faculty told us that although teaching was supposedly important, publishing was the key to getting tenure and where we should concentrate our energies. What counted as publications was not always clear either. For example, at two schools, some faculty suggested that research in progress and books counted towards tenure, while others said no.

When department members present conflicting signals, our experiences suggest gathering information about the issue from a variety of other sources at each institution and discussing the information with a variety of faculty and administrators. For example, with tenure issues, we recommend seeing if any written guidelines exist. If so, examine them, and then discuss the conflicting information with the chair of the department and/or the dean.

Requesting Candidate Information. At several schools (n = 7) faculty members asked us questions of a sensitive or confidential nature. As sociologists, many of whom study issues related to inequality, we knew these questions were off limits for interviewers to ask. We found that many interviewers were either not aware or respectful of these limits.

Faculty often asked us about our current or previous graduate department. We were wary of these questions, particularly if someone mentioned specific things about the department or individual faculty members. A candidate never knows whether the interviewer may be a friend or foe of someone about whom they are inquiring. We tried to speak positively about others, as we knew that speaking negatively of others might reflect poorly on us.

Legally, interviewers should not ask sensitive demographic information such as marital or parental status. Often interviewers can determine this information a priori from their own networks. Or, they determine it in casual conversation with the candidate during the interview. Even though it was not always meant to be malicious, we were still surprised at the frequency with which we received direct questions about our marital status. Many times, these questions came from members of the department and even from members of the search committee (who we presume should know what questions can and can not be asked).

Candidates should weigh the potential advantages and disadvantages of disclosing sensitive information. Disclosure may affect faculty members' perceptions of candidates, in at least two ways. On the one hand, departments may place a lower probability of a candidate accepting an offer if he or she is married or partnered. They may also assign a lower probability of retention if the candidate's spouse or partner does not get a job in the area.

On the other hand, some departments may view a candidate with a partner and children as a predictor of possible employment retention. Once settled, it is more difficult for families than individuals to relocate. For two of our co-authors, partner's employment in the local labor market was important to the acceptance of a job offer. Hence, one of these co-authors found it important to discuss local employment opportunities with chairs. A couple of us also found that some institutions contain offices established to help with partner relocation.

In sum, our experiences suggest that departments' presentation of self varies according to the type of institution, gender make-up of the department, and the school's prestige. Our experiences also suggest that departments may want to work on clarifying and unifying their individual and collective presentations of self. While this definition of the situation represents the co-authors' experiences, we want to reiterate that it may not necessarily represent all institutions.

Administration Presentation of Self

During our interviews we often met with a Dean or Assistant Dean, about half of whom were men. These interviews provided the opportunity for us to assess how administrators presented themselves, the department, and the university. While our time with administrators was limited, we immediately noticed a more formalized presentation than that of departmental members. Most deans used a business style persona, regardless of gender. All dressed professionally and referred to us with a title. Nearly all of them spoke of the larger vision of the university and most mentioned how the sociology department fit into that vision. They also often conveyed a stronger interest in our external funding capabilities than in our teaching abilities.

Some of our experiences indicate that administrators were fairly honest in their assessments of departments. For instance, one administrator noted that the sociology department was one of the lowest in terms of external research funding. Two, however, failed to disclose internal frictions in a department that resulted in various legal actions. Both candidates later found out about these matters through an inside source (e.g., a visiting instructor who worked in the department). Administrators at other schools (n = 4) professed to special programs such as applied or international programs, but really appeared to only "pay lip service" to these programs, never describing the history, details, or future plans of these programs.

At a few schools (n = 4) administrators discussed opportunities for growth and change in the sociology department. For example, several departments were "top heavy" with a lot of older, full professors, but few if any junior faculty. These administrators emphasized the changing nature of the department and the opportunities a new faculty member would have in helping to shape the future of the department.

All in all, administrators presented themselves more as a resource to connect us to the larger university. They could cite average SAT scores of new students, retention rates, most popular majors, socio-demographics of their student body, and assorted statistics associated with the school. One of our co-authors realized this fairly early in the process and proceeded to prepare a variety of questions for these meetings. In essence, she took control of these interviews and attempted to discern as much about the department and university as possible.

The ASA and Job Searching

Although few resources detailing aspects of job searching in sociology exist, the ASA does provide a variety of assistance for job seekers. One of the primary ASA products used by many sociologists seeking positions is the *Employment Bulletin*, which identifies academic job openings, mostly in the U.S. Most of our five co-authors relied heavily on the ASA's *Employment Bulletin*, pouring over its contents each month. One of the advantages of this resource is that the ASA offers it free each month via its website (*www.asanet.org*). Hard copies of the *Employment Bulletin* may be ordered, for a fee, from the ASA. The ASA also publishes *Research Briefs* and short books providing aggregate information on Ph.D. production, job characteristics, and demographics in the sociology academic department (Spalter-Roth et al. 2000; Spalter-Roth and Lee 2000).

The ASA usually provides several sessions for professional socialization among job seekers at the annual national meeting. These sessions detail certain aspects of job searching in sociology and in other, more applied, fields. Students, in particular, benefit from these sessions because they can learn about the experiences of others in the academic labor market, including the perspective of applicants and search committee chairs.

Another valuable resource provided by the ASA is the Employment Center held during the annual national meeting. For a fee, the Employment Center allows both candidates and school representatives to review information about and request interviews with each other. Departments post advertisements about their positions and registered job seekers place their vitas on file for prospective employers to review. Employers who can not attend the meeting can purchase job seekers' vitas. Other national and regional sociological conferences such as the American Society of Criminology, the Southern Sociological Society, and the Eastern Sociological Society also provide employment centers and/or professional socialization sessions. Levine (2000) notes that the ASA Employment Center is one of the first ways candidates usually test the academic job market.

While our experiences suggest that interviews conducted at the ASA Employment Center seldom result in job offers, they do serve several beneficial purposes. First, they allow candidates the opportunity to interview with schools in a somewhat more relaxed environment than during a campus interview. This allows candidates the opportunity to become familiar with the types of questions interviewers ask and what information they need to provide. It also allows candidates the opportunity to hone their presentation of self in these types of situations.

At a basic level, the Employment Center provides a venue through which job seekers and departments can find out more information about each other than through traditional job advertisements, cover letters, and vitas. This may result in candidates becoming interested in schools they would not have been interested in previously. Similarly, information gained from an ASA Employment Center interview may supplement that provided from job applicants' vitas and cover letters.

However, one of the things that surprised us about the employment process at professional meetings was the "meat market" feel of it. At the ASA Employment Center, we sat in chairs awaiting the ring of a bell that would signal the end of one interview and the beginning of another interview. We often "chafed at the bit," waiting for someone to leave the table from the school where we were next scheduled to interview. Should we approach the table and assert our status as the next candidate, or wait until the other candidate left? Although we initially often waited, we soon learned that if we didn't proceed to announce ourselves the candidate and interviewer(s) would often continue well into our allotted interview time.

We definitely learned that time was an issue in this interview process, and that we had to assert our right to claim our interview time while still managing the impression we were presenting. Whether these patterns exist within other disciplines, we do not know. Given what we experienced, we suggest that sociologists who interview candidates in Employment Centers adhere to time allotments for each candidate and note the importance of keeping on schedule during these interviews.

Although the time allotted for sociology meetings is limited and attendees generally have many scheduled commitments, we were surprised at the number of candidates that some schools saw during the meetings, and even some during one day. Some interviewers saw more than 25 candidates in a given day. With many departments seeing large numbers of candidates in a given day, we wondered how they could remember who said what or who had what qualifications. Knowing this, candidates should definitely try to differentiate themselves during Employment Center interviews.

Given our experiences, it seems reasonable to think that sociology employment services should alter the time limits for these interviews. Perhaps 30-45 minutes would provide a better opportunity for schools to discover whether they are really interested in candidates, and perhaps schools would screen candidates more carefully to ensure a better fit with their needs. We realize, however, that this might place greater time constraints on already taxed faculty members' time. More radically, given that most schools, due to structural constraints, are not able to advertise positions until after September (Kronenfeld and Whicker 1997), we question the effectiveness of holding an Employment Center in mid August.

An unanticipated outcome, for us, of interviewing at professional meetings was the opportunity to meet others that were going through the same process and possibly applying for the same jobs as us. Although there was some ambivalence about finding out whom else was interviewing in our specialty areas and their qualifications, several of us made contacts with other candidates at these conferences. Two of our co-authors initially met this way. They exchanged contact information, saw each other at other sociology conferences, and continued to exchange advice and information about the job search process (and the peculiarities associated with it) within sociology.

Besides participating in Employment Centers, attending the ASA and other sociological conferences provides an opportunity to make contacts with faculty in departments where job openings exist and places where candidates plan to apply. Given the number of applications departments receive, being able to link a name to a face may help candidates stand out among the other candidates, for better or worse depending upon the quality of the self-presentation. The meeting program typically provides the names of participants and the sessions in which they will be presenting. Several of our co-authors found it useful to attend those sessions not only to meet the person, but also to learn about their research. From our perspective, it was never too early to start gathering information about an institution. Even if we didn't receive campus visit requests, we still made good contacts that might be useful in the future.

The resources provided by the ASA certainly help sociology job seekers to find positions and to become better at their self-presentation. However, we suggest that this is not enough. The ASA needs to publish a guide to getting an academic job in sociology. This publication would provide a wealth of details that would be beneficial to job seekers. Currently, students mainly learn about the processes involved in getting an academic sociology job through mentors, colleagues, and their own experiences.

For instance, do candidates know that searching for academic jobs is one of the most time-consuming aspects of job searching? Do they know that employers in the sociology labor market generally adhere to a timeline concerning the review of applications, scheduling and conducting campus interviews, and making job offers? Do candidates know that most schools advertise their positions in early to late fall, with applications due by late November to mid December? Do they know that most jobs are not filled via the ASA Employment Center? And, do they know that most positions advertised in the early months of the following year are often non-tenure track positions? A publication that provides details about the wealth of information associated with the sociology academic labor market would be an invaluable service to both job seekers and employers.

Conclusion

Through this experience, we learned a great deal about sociology as a discipline, the hiring practices within the discipline, and identity work and self-presentation among job candidates and schools. Some of our experiences indicate that the hiring practices within sociology are inefficient. To select candidates, we rely mainly on impressions garnered through cover letters, vitas, references, and professional networks. Then, we have limited funds to bring candidates to our campuses. Usually, we give job candidates very little time to learn about the schools with which they interview, often scheduling interviews within two weeks from invitation. We place an extreme emphasis on the campus interview—for both the candidate and the department. With all of our collective "eggs in one basket," the campus interview becomes very stressful for everyone involved.

Many of these problems are a manifestation of how our labor market is structured. It allows usually one opportunity—the campus interview—for applicants and departments to meet. Thus, the presentations during this meeting become extremely important for both applicants and departments. Restructuring our job search processes could provide additional opportunities for applicants and departments to meet. For example, we could follow the way in which economists organize their job search process. They advertise most positions in October and November, with deadlines of early December (Carson and Navarro 1988). Then, job search committees meet to review applicants in early December and invite a relatively large list of selected candidates to interview with them at their national meetings in early January. From these, they choose a handful of candidates to invite for campus visits.

We could follow this model by holding our national meetings in January or early February. As it is now, our national meetings occur before most job advertisements have even been written, let alone advertised. Moving the annual meeting would allow more candidates and departments to meet, gather information about each other, and, theoretically, make better choices about who to invite to campus and which invitations to accept. Better choices could save applicants and departments time, money, and work.

At a more cultural level, we suggest that the discipline reconstruct the goal and content of rejection letters sent to academic job candidates. Shaw (2000) suggests that rejection letters help us to develop a sense of reality about our discipline. If this is true, then our rejection letters paint an ugly picture of sociology. Most of our rejection letters were terse letters with almost no information on the hiring process, other than that "a suitable or more qualified candidate was hired." And, we never received any type of letter—rejection or otherwise—from many schools to which we applied.

Following Shaw's (2000) recommendation, we suggest that after all applications are reviewed, departments prepare letters for candidates that provide details on (1) the total number and quality of applications received, (2) the selection process that the search committee will use, and (3) the timeline associated with the selection process. Final rejection letters should detail the credentials of the person hired for the position.

We are not suggesting that departments, specifically search committee chairs, write an individual rejection letter for each candidate. Rather, we simply advocate providing candidates with more details of the whole process and the qualifications of the candidate hired for the position. Providing more information on the job search process and outcomes will aid job candidates in determining why they were not selected for the position, and hopefully help them improve their future job searches.

Finally, we question the heavy weight that we place, as a discipline, on job talks. During a job talk, we expect candidates to present specific information about their present and future research. As such, self-presentation in job talks usually reveals little about a candidate's knowledge of the discipline of sociology as a whole. Meanwhile, from a campus visit, most departments seek to gain a general assessment of a candidate. Given that most departments do not request candidates to teach a class, the job talk becomes the main event of most campus interviews. With the job talk revealing specific information and the departments seeking general information, have we, as a discipline, created a process that routinely fails to effectively match candidates and departments? Future research in this area will hopefully help to answer this question.

As we stated earlier, our experiences may not be reflective of all individuals who have moved into assistant professor positions. Given that we entered the mid-tier labor market, the processes and outcomes for the labor market in top 20 schools may differ (for a discussion of this point in economics see Carson and Navarro 1988). However, we hope that our experiences and the lessons we learned will illuminate some of the social psychological aspects associated with job searching in sociology and some of the lessons we learned about the discipline of sociology along the way. Perhaps it will also help future sociology job seekers as they enter the academic labor market. Reading this article should at least help them identify some of the issues they may face.

References

Aguirre, Adalberto. 2000. "Academic Storytelling: A Critical Race Theory Story of Affirmative Action." *Sociological Perspectives* 43(2): 319-339.

Ault, Brian. 1996. "Structure of Graduate Student Failure: A View from Within," *The American Sociologist* 27(3): 27-38.

Boufis, Christina and Victoria C. Olsen (Eds). 1997. *On the Market: Surviving the Academic Job Search*. New York: Riverhead Books.

Brems, Christiane, Claudia Lampman, and Mark E. Johnson. 1995. "Preparation of Applications for Academic Positions in Psychology," *American Psychologist* 50(7): 533-537.

Brewer, Marilynn B. 1991. "The Social Self: On Being the Same and Different at the Same Time," *Personality and Social Psychology Bulletin* 17(5): 475-482.

Broughton, Walter. 2000. "Strengthening the Market for Sociology Ph.D.s," *Eastern Sociological Society Newsletter* 14(4): 3-4.

Carson, Richard and Peter Navarro. 1988. "A Seller's (& Buyer's) Guide to the Job Market for Beginning Academic Economists," *Journal of Economic Perspectives* 2(2): 137-148.

Chronicle of Higher Education (no author listed). 1992. "Fruitless Interviews and Tasteless Rejections: Job Hunting in Academe," *Chronicle of Higher Education* 39(15): B3.

Chronicle of Higher Education (no author listed). 1994. "Tough Times in the Academic Job Market," *Chronicle of Higher Education* 40(34): A1.

Cohen, Stanley. 1997. "Conference Life: The Rough Guide," *The American Sociologist* 28(3): 69-84.

Darley, John M. and Mark P. Zanna. 1987. "The Hiring Process in Academia." Pp. 3-21 in Zanna and Darley (Eds.), *The Compleat Academic*. Mahwah, NJ: Lawrence Erlbaum.

Deegan, Mary Jo. 1991. *Women in Sociology: A Biographical Sourcebook*. New York: Greenwood Press.

Gecas, Viktor. 1982. "The Self-Concept," *Annual Review of Sociology* 8: 1-33.

Goetting, Ann and Sarah Fenstermaker. 1995. *Individual Voices, Collective Visions, Fifty Years of Women in Sociology*. Philadelphia: Temple University Press.

Grimes, Michael D. and Joan M. Morris. 1997. *Caught in the Middle: Contradictions of Sociologists from Working-Class Backgrounds*. Westport, CT: Praeger.

Heiberger, Mary Morris and Julia Miller Vick. 1992. *The Academic Job Search Handbook*. Philadelphia: University of Pennsylvania Press.

Hochschild, Arlie R. 1979. "Emotion Work, Feeling Rules and Social Structure," *American Journal of Sociology* 85(3): 551-575.

Hotaling, Debra. 1994. "New Ph.D.s Can Find Life Outside Academe," *Chronicle of Higher Education* 41(7): B1.

Kahn, Shulmait. 1993. "Gender Differences in Academic Career Paths of Economists," *The American Economic Review* 83: 52-56.

Kronenfeld, Jennie Jacobs and Marcia Lynn Whicker. 1997. *Getting an Academic Job: Strategies for Success. Survival Skills for Scholars*, Volume 17. Thousand Oaks, CA: Sage.

Levine, Felice. 2000. "Sociologists at Work," *Footnotes* 28(9): 2.

Magner, Dennis K. 1997. "Job Market for Ph.D.s Shows First Sign of Improvement," *Chronicle of Higher Education* 43(21): 102-125.

Marx, Gary T. 1997. "Of Methods and Manners for Aspiring Sociologists: 37 Moral Imperatives," *The American Sociologist* 28(1): 102-125.

Misra, Joya, Ivy Kennelly, and Marina Karides. 1999. "Employment Chances in the Academic Job Market in Sociology: Do Race and Gender Matter?" *Sociological Perspectives* 42(2): 215-247.

Orlans, Kathryn, P. Meadow, and Ruth Wallace. 1975. *Gender and the Academic Experience: Berkeley Women Sociologists*. Lincoln, NE: University of Nebraska Press.

Rheingold, Harriet L. 1994. *The Psychologist's Guide to an Academic Career*. American Psychological Association

Sawyer, R. McLaran, Keith W. Prichard, and Karl D. Hostetler (Eds). 1992. *The Art and Politics of College Teaching: A Practical Guide for the Beginning Professor*. New York: Lang Publishing.

Siegfried, John J. and Wendy A. Stock. 1999. "The Labor Market for Ph.D. Economists," *Journal of Economic Perspectives* 13(3): 115-134.

Shaw, Victor N. 2000. "Toward Professional Civility: An Analysis of Rejection Letters from Sociology Departments," *The American Sociologist* 31(1): 32-43.

Sheehan, Eugene P. and Hollis Haselhorst. 1999. "A Profile of Applicants for an Academic Position in Social Psychology," *Journal of Social Behavior & Personality* 14(1): 23-31.

Shields, Stephanie A. 1987. "Women, Men, and the Dilemma of Emotion." Pp. 229-250 in P. Shaver and C. Hendrick (Eds.), *Sex and Gender*. Newbury Park, CA: Sage.

Shively, Gerald, Richard Woodward, and Denise Stanley. 1999. "Strategy and Etiquette for Graduate Students Entering the Academic Job Market," *Review of Agricultural Economics* 21(2): 513-526.

Spalter-Roth, Roberta and Sunhwa Lee. 2000. "Gender in the Early Stages of the Sociological Career." American Sociological Association Research Program on the Discipline and Profession, Research Brief.

Spalter-Roth, Roberta, Jan Thomas, and Felice J. Levine. 2000. "New Doctorates in Sociology: Professions Inside and Outside the Academy." American Sociological Association Research Program on the Discipline and Profession, Research Brief.

Stith, Sandra M. 1990. "Preparing for a Research-Focused Career in Family Studies," *Family Relations* 39: 456-459.

Weiss, Robert. 1990. *Staying the Course: The Emotional and Social Lives of Men Who Do Well at Work*. New York: Free Press.

Zanna Mark P. and John M. Darley. 1987. *The Compleat Academic*. New York: Random House.

Article

Social Capital Activation and Job Searching: Embedding the Use of Weak Ties in the American Institutional Context

Work and Occupations
2014, Vol. 41(4) 409–439
© The Author(s) 2014
Reprints and permissions:
sagepub.com/journalsPermissions.nav
DOI: 10.1177/0730888414538432
wox.sagepub.com

Ofer Sharone[1]

Abstract
By comparing job seekers' use of weak ties in Israel and the United States, this article shows that Granovetter's canonical findings are rooted in the particular institutional context of the American white-collar labor market. Drawing on in-depth interviews with three distinct groups of white-collar job seekers: Americans searching in the United States, Israelis searching in Israel, and Israelis searching in the United States, this article untangles cultural and institutional factors underlying the use of weak ties and shows how labor market institutions and processes of hiring shape systematic variations in job seekers' utilization of weak ties.

Keywords
weak ties, social capital, job searching, qualitative methods, cross-national comparison, culture

---

[1]Work and Organization Studies, Massachusetts Institute of Technology, Sloan School of Management, Cambridge, MA, USA

Corresponding Author:
Ofer Sharone, Work and Organization Studies, Massachusetts Institute of Technology, Sloan School of Management, 100 Main Street, Bldg. E62-340, Cambridge, MA 02142, USA.
Email: osharone@mit.edu

The use of personal contacts to find work is an age-old phenomenon. A 1930s study found that more than half of the workers at the hosiery factories of Philadelphia had found their jobs through personal ties (De Schweinetz, 1932). This method of finding a job is not limited to any particular class or occupation. No less a talent than Albert Einstein relied on whom he knew rather than what he knew to obtain his first professional job (De Graaf & Flap, 1988).

The use of ties among white-collar workers began to receive intense scholarly attention with Granovetter's (1973, 1974/1995) groundbreaking research showing the significant extent to which American white-collar workers use social ties to find jobs and, counterintuitively, finding that more workers found their jobs through weak rather than strong ties. Granovetter theorized that weak ties were more effective than strong ties because they provide nonredundant information about job openings. Although many scholars find Granovetter's explanation compelling, studies examining the efficacy of weak ties outside the United States have yielded mixed results (e.g., Bian, 1997; Yakubovic, 2005).

The existing literature's ongoing debate on the efficacy of weak ties leaves largely unexamined an equally important question regarding the activation of weak ties. Specifically, under what institutional conditions do actors, such as job seekers, attempt to utilize weak ties? The dearth of studies on this issue leaves unclear the institutional underpinnings for Granovetter's (1973, 1974/1995) findings regarding the extensive utilization of weak ties among American white-collar job seekers and, more broadly, the institutional conditions that facilitate or inhibit efforts to activate weak ties. The absence of comparative qualitative data leaves actors' understandings and strategies regarding the use of weak ties in a black box and therefore leaves the existing literature unable to effectively account for systematic variations in the activation of weak ties.

This article opens this black box by comparing white-collar job seekers' understandings and practices in activating weak ties to find work in two advanced economies. Drawing on in-depth interviews with three distinct groups of white-collar job seekers—Americans searching for work in the United States, Israelis searching in Israel, and Israelis searching in the United States—this article examines the factors underlying the activation of weak ties and shows the hitherto unrecognized role of labor market institutions and processes of hiring in accounting for systematic variations. Specifically, this article links the well-documented efforts of American white-collar job seekers to activate weak ties (Lane, 2011; Sharone, 2013b; Smith, 2001) to the particular institutions structuring the American white-collar labor market. More broadly, by examining the link between labor market institutions and job seekers' utilization of ties, this article contributes to our understanding of the interrelationship between institutional structures and agents' activation of social networks.

Weak-Tie Activation: Granovetter and Theories of Variations

In Granovetter's study (1973, 1974/1995), white-collar workers who found their jobs through personal contacts used weak ties more often than strong ties. Granovetter's canonical explanation for this finding focuses on the distinct usefulness of information obtained from weak ties. As Granovetter (1973) put it: "Those to whom we are weakly tied are more likely to move in circles different from our own and will thus have access to information different from that which we receive" (p. 1371). It is the access to new information about job opportunities that makes weak ties strong.

The existing literature has focused on the efficacy of using weak ties across social contexts and reveals mixed results. Yakubovic (2005), studying the Russian labor market, finds support for Granovetter's (1973, 1974/1995) thesis on the efficacy of weak ties, whereas Bian (1997) finds that strong ties were more useful in 1980s China. Other studies find that the strength of ties is not significant (e.g., Korpi, 2001; Marsden & Hulbert, 1988). In explaining these variations, the literature has focused on how contextual factors shape the usefulness of information obtained from social ties. For example, Burt (1977) finds that social ties are more valuable sources of information for managers with few peers than for managers with many peers. Other studies have focused on the variations arising from the nature of the information to be transmitted by social ties. Aral and Alstyne (2011) find that if one needs *thick* information in an environment where there is rapid change, strong ties are better able to provide such information than weak ties. Pointing to a trade-off between "diversity" and "bandwidth," Aral and Alstyne find that while "the prevailing wisdom among sociologists for the last 40 years has been that the strength of weak ties and informational advantages to brokerage operate with a fair degree of regularity across contexts, our analysis shows that context matters" (p. 148). Specifically, in contexts with more "information turbulence," strong ties—which come with greater bandwidth—may be more valuable than weak ties with their more diverse forms of information (Aral & Alstyne, 2011, p. 148; Reagans & McEvily, 2003).

Another set of studies claims that cross-national variations with respect to the efficacy of ties are attributable to cultural factors. Pointing out that most prior research on the use of weak ties has been conducted in the West, these studies argue that the literature fails to sufficiently consider the effects of culture—specifically, cultural contexts characterized by *collectivism* rather than *individualism* (Ma, Huang, & Shenkar, 2011; Xiao & Tsui, 2007). Utilizing conceptions of culture rooted in the management literature (e.g., Hofstede, 2001), which are challenged by many contemporary sociologists (e.g., Swidler, 1986), these articles argue that collectivist cultural contexts entail high levels of cooperation among one's *in group* and hence make the use of strong ties and closed networks effective, while the same cultural context entails comparatively low levels of cooperation with members from out groups and thus renders less effective the utilization of weak ties or structural holes.

Shifting the focus from abstract cultural contexts to concrete institutions, Bian (1997) examines the role of labor market practices in 1980s China—in the era, predating market reforms, when state officials allocated jobs. Bian finds that in the institutional context of state socialism, strong ties are more effective than weak ties because helping one's ties in the hiring process requires state officials to engage in risky unauthorized activity. Given these risks, job seekers could rely on only strong ties—or the strong ties of strong ties—who would be highly motivated to help them on the basis of a past relationship of trust and mutual obligation. Although Bian provides a compelling explanation for the reliance on strong ties in a socialist context given its very distinct hiring process, it is not clear to what extent these findings may apply to white-collar workers in advanced market economies.

In sum, prior studies focus on a host of contextual factors that may shape the efficacy of weak versus strong ties. However, to date, less attention has been paid to how contextual factors shape the activation of weak ties. Although Bian (1997) focuses on the efficacy of weak versus strong ties, his focus on how contexts may shape the motivation of ties to help provides an important point of departure for considering mechanisms that may explain variations in the activation of weak ties. As the social capital literature has recently emphasized, to account for the use of weak ties to find jobs, our research focus must expand beyond the potential usefulness of the information obtained from such ties and address the question of motivation (Marin, 2012; Obukhova, 2012). Job seekers are likely to invest time and energy in attempting to activate weak ties only if they perceive some sources of motivation for such weak ties to help them.

The unclear motivation of weak ties was recognized by Granovetter (1974/1995) and was one of the reasons that his findings about the prevalent use of weak ties were considered counterintuitive. As Granovetter (1973) put it, "a natural a priori idea is that those with whom one has strong ties are more motivated to help with job information" (p. 1371). The motivation of weak ties is particularly puzzling when we realize just how weak some such ties are. In Granovetter's (1974/1995, p. 81) study, some weak ties provided useful information even in cases where the ties "had barely been maintained at all" for periods as long as 20 years. Granovetter (1973) himself noted that it is a remarkable fact that "people receive crucial information from individuals whose very existence they have forgotten" (p. 1372).

Why would a forgotten weak tie pass on valuable information? The question can be asked even more pointedly in light of Granovetter's (1974/1995, p. 28) finding that the weakest of the weak ties, those who had only "rarely" been in touch, also tended to put in a "good word" about the job seeker to the employer and thus, to some extent, put themselves on the line. What would motivate weak ties whose *existence one has nearly forgotten* both to pass on valuable information about job openings and to put in a *good word* with the employer?

Granovetter (1974/1995, p. 54) offers an explanation for why weak ties—who, unlike strong ties, are "not likely to be under any particular pressure" to help their job-seeking contacts—might nonetheless do so. He claims that "those who are able to recruit competent personnel may find their reputations enhanced; they will appear to be people who know how to get things done" (p. 55). A recent study of the referring behavior of insurance agents in Toronto lends support to Granovetter's claim, finding that white-collar workers refer weak ties not only because it is the "obvious and nice thing to do" but also because they see such referrals as opportunities to enhance their reputations (Marin, 2012, p. 186).

The literature's understanding of the motivation of weak ties is premised on the notion that a successful match yields a win–win–win outcome. The referee gets valuable help in his or her job search, the employer gets a good new employee, and the referrer enhances his or her reputation with the employer. The apparent benefits that flow to all involved leads Granovetter (1974/1995) to wonder about the extent to which the mechanisms that motivate the activation and responsiveness of weak ties in the American white-collar labor market would also yield similar outcomes in different social contexts. When considering whether his findings are "peculiar to higher white-collar workers in Massachusetts" or are "aspects of the human condition," Granovetter

(1974/1995, p. 119) acknowledges that there are very few existing data on this subject outside the American context, but sees "no reason to expect processes in industrial non-American markets... to be radically different from those reported here" (p. 130). Cautiously limiting his statement about potential generalizability to a double negative, he notes that "it is not clear that my results do not reflect the situation in a wide variety of economies" (Granovetter, 1974/1995, p.130). He is explicit that "positive conclusions must await further research" (p. 130). This article begins to provide that much-needed further research.

No comparative research to date has qualitatively explored cross-national variations in job seekers' efforts to activate weak ties or how contextual factors shape job seekers' understandings of labor-market processes—most critically their understandings of the motivations of weak ties—and consequently their strategies in activating weak ties. The existing literature acknowledges this gap. Granovetter (1974/1995), in the *Afterward* to the second edition of his book, notes the continued scarcity of comparative data. This lack of data has also been observed by Gerber and Mayorova (2010) and Xiao and Tsui (2007).

This article compares the use of weak ties by white-collar job seekers in two advanced market economies and shows that the activation of weak ties systematically varies across these different economies. It finds that job seekers looking for work in the United States, whether they are culturally American or Israeli, are much more likely than job seekers in Israel to use weak ties. Moreover, this article shows that whether or not job seekers attempt to use weak ties to find a job depends on how institutional contexts shape their understandings of labor markets and the motivations of weak ties and employers. The mechanisms Granovetter (1974/1995) describes in explaining the use of weak ties in the United States are not universal but vary with job seekers' understandings of the workings of the hiring process.

Three-Way Comparison: Americans, Israelis, and Israelis in the United States

To compare the use of social ties to find jobs across institutional contexts, this article examines the job search practices of three groups of unemployed white-collar workers. The first group, whom I will refer to as *Americans*, are job seekers who are looking for work in the San Francisco and Boston areas and who have lived in the United States for at least 10 years. The second group, whom I will refer to as *Israelis*, are job seekers looking for work in the Tel Aviv area and who have lived in Israel for at least 10 years. The third group, whom I will refer to as *Israelis in the United States*, are job seekers who are looking for work in the Boston area and who have arrived in the United States from Israel less than one year before the interview. By including in the sample Israelis looking for work in the United States and comparing them with Americans looking in the United States and Israelis looking in Israel, this article is able to move the analysis of the activation of weak ties beyond internalized cultural norms of individualism or collectivism (Ma et al., 2011; Xiao & Tsui, 2007) and to instead focus on the effects of institutional contexts.

Although Israel and the United States have significant differences, such as their respective economic sizes (which will be discussed later), the comparison of white-collar job seekers in the San Francisco, Tel Aviv, and Boston areas is particularly interesting because of a number of important similarities across these metropolitan areas. All three cities rank among the most important high-tech, biotech, and start-up hubs in the world (e.g., Rosenberg & Vainunska, 2007).[1] For example, a recent industry analysis ranks Silicon Valley, Tel Aviv, and Boston first, second, and sixth, respectively, in the world for start-ups.[2] Beyond these industry-level similarities, over the past 30 years white-collar workers in both Israel and the United States have experienced a sharp rise in job insecurity (Kalleberg, 2009, 2011; Ram, 2008; Samuel & Harpaz, 2004) due to labor-market deregulation in Israel and changing corporate practices in the United States that have made layoffs of white-collar workers routine in both countries (Mishel, Bernstein, & Allegretto, 2007; Osterman, 1999; Samuel & Harpaz, 2004).[3] Finally, on becoming unemployed, workers in both countries face similar regimes of unemployment insurance (Gal, 2005; Gangl, 2004; Hipp, 2011). Given these similarities, we may expect similar job-searching practices and efforts to activate weak ties.

To explore the activation of weak ties in the course of job searching, I conducted in-depth interviews with 84 Americans from the San Francisco and Boston areas, 48 Israelis from the Tel Aviv area, and 20 Israelis searching for work in the Boston area.[4] Each interview lasted between 90 min and 3 hours. The interviews were semistructured and asked open-ended questions about job search experiences and strategies. A portion of each interview asked job seekers to describe any attempts they had made to activate social networks. I limited my interviews to unemployed job seekers[5] in the middle years of their working lives, between the ages of 25 and 65. To facilitate comparability, I also limited my interviews to white-collar workers looking for work in the private sector. General occupational categories were also kept comparable across sites, with roughly similar proportions of managers, technical workers, professionals, and other white-collar workers.[6] Table 1 summarizes the basic demographic characteristics of the interviewees.

Approximately half of both the American job seekers in Boston and San Francisco and the Israeli job seekers in Tel Aviv were recruited at support groups for white-collar job seekers.[7] Another 40% of these interviewees were recruited by randomly approaching individuals at government unemployment offices and job fairs. Finally, 10% were recruited by snowballing. In addition to interviews, I observed support groups for white-collar job seekers—one in the San Francisco Bay area, which I will refer to as *AmeriSupport*, and one in the Tel Aviv area, which I will refer to as *IsraSupport*.

Regarding the Israeli job seekers in Boston, 60% were recruited at workshops aimed at helping recently arrived Israelis look for work and 40% were recruited by snowballing. It is important to note that more than half of these Boston-based Israeli job seekers were spouses of postdoctoral scholars at one of the Boston-area universities. Thus, unlike typical immigrants who self-select to seek work in the United States,

Table 1. Sample Demographics.

Job seekers	Americans in San Francisco and Boston areas	Israelis in Tel Aviv area	Israelis in Boston area
Number of interviewees	84	48	20
Percent men	54	58	30
Age (median)	46	43	31
Percent married	45	56	80
Over one year unemployed (%)	40	36	n/a
Race/ethnicity (%)			
White	79		
African American	9		
Hispanic	3		
Asian American and other	9		
Jewish European descent	n/a	48	60
Jewish Middle Eastern descent	n/a	28	20
Jewish Other	n/a	24	20

these job seekers share only a willingness to accommodate their partners' careers. My interview data for this group of job seekers were supplemented by observations of the workshops supporting newly arriving Israelis in Boston.

Comparing Networking Practices in the United States and Israel

My interview data reveal that job seekers in the United States and Israel have sharply diverging understandings and practices regarding the use of ties. Job seekers in the United States place at the center of their job search the activation of all ties, strong and weak, as well as the cultivation of new ties—a form of tie not addressed by Granovetter (1974/1995) but which, as discussed later, is pursued by American job seekers with the same logic as the activation of weak ties. In contrast, job seekers in Israel turn only to strong ties—close friends and relatives—for help in their job search and do so only after they have exhausted other search methods.

Nearly all the American white-collar job seekers whom I interviewed believed that *networking* was critical to finding work, and most made strenuous efforts both to network with existing strong and weak ties and to generate new ones. The day-to-day networking practices of American white-collar job seekers are primarily aimed at weak ties or the creation of new ties because job seekers typically turn to their strong ties at the start of their search and quickly exhaust job leads arising from this relatively small pool of ties. David, an American high-tech manager, explained the conventional wisdom: "The way you will get a job is through people that you know." To get to know more people, he uses online social networks and attends conferences and networking events put on by high-tech companies. He points to a flash disk that he wears around his neck and proclaims: "I have 5,000 contacts right here!" Frank, likewise, explained his focus on networking: "How do I look for work? Networking is the best way. When someone gets the idea to hire someone, they first ask their friends." Many American job seekers I interviewed used the same expression, explaining that to find work you need "to put yourself out there." The intense focus of American white-collar job seekers on rekindling old ties and cultivating new ties is found by all recent studies of white-collar job searching in the United States (e.g., Ehrenreich, 2005; Lane, 2011; Sharone, 2013b; Smith, 2001).

While Granovetter (1974/1995) found implausible the idea that efforts to cultivate new contacts would be effective,[8] American job seekers who regularly attend conferences and networking events hoping to make new contacts clearly believe that this is effective, as do university career offices, professional associations, and state-run one-stop centers, which routinely organize such networking events (Ehrenreich, 2005; Lane, 2011; Smith, 2001). Given the centrality of these practices to contemporary white-collar job searching in the United States, this article will explore both job seeker practices that are aimed at generating new ties, as would be the case at networking events (Lane, 2011), as well as activating a preexisting weak tie, such as a former colleague. Indeed, given Granovetter's (1973, 1974/1995) previously discussed theory of the motivations of referrers of weak ties to enhance their reputations as people who "know how to get things done," it appears reasonable to expect that efforts to create new ties may be as effective as cases documented by Granovetter, where job seekers contacted weak ties with whom they had not spoken for 20 years (p. 55).

In contrast to job seekers in the United States, job seekers in Israel seek help only from well-established strong ties and do not attempt to mobilize weak ties or expand their existing networks. Strikingly, I did not find a single exception to this pattern in Israel. Israelis who were job searching in the Boston area (even in the very town where Granovetter (1973, 1974/1995) conducted his research nearly 40 years ago) and who had previously searched for work in Israel most clearly articulated the different approaches to weak ties by drawing on their firsthand experiences with both types of networking practices. For example, in contrast to Don's current networking practice in the Boston area, which is focused on identifying weak ties at potential employers, during his most recent search in Israel, he chose not to reach out to an acquaintance working at a company where he interviewed. He reflected: "In my last job in Israel, when I interviewed, I knew someone at the company but I did not contact him. In the United States, I would have contacted him."

Dorit, an Israeli manager looking for work in Boston, also takes a different approach to using social ties in the United States than she did in Israel. She noted the differences with respect to trying to create new contacts through introductions:

> Networking here does not look at all like in Israel... Here there is the practice of "introduction"—people introduce each other very easily. In Israel, this is not done... There is no concept of "introduction." In Israel, they help if they are a close friend, but someone not close will not help.

Israeli job seekers, when looking for work in Israel, dismissed the idea of attempting to create new ties. Eldad, a marketer, told me: "Personal connections are important," but he added, "if they don't know you, forget it." Or as Don put it: "In Israel, [networking] is not active. Personal contacts may help, but it is not contacts that were formed during the job search." The characterization of Israeli networking as *not active* also points to the fact that, unlike job seekers in the United States, job seekers in Israel make no effort to reconnect with old colleagues or to identify friends of friends who may know a hiring manager at a targeted employer.

Hila, an Israeli career coach who guides Israelis looking for work in Boston, put the difference this way:

> In Israel, we do not say that a worker brings a friend but that a friend brings a friend. The basis [for a referral] is ultimately social in Israel. You will not refer someone you do not socially know. People will only help those to whom they are socially close.

The social basis for help in Israel rules out the use of the most common weak ties activated by white-collar American job seekers—former work colleagues whom they never saw outside the work context (Granovetter, 1973).

One clear indication of the difference between the Israeli and American practices of networking with weak or new ties is seen with respect to the prevalence of so-called *networking events*. Gatherings set up for job seekers to meet and network are ubiquitous in the United States, and most American job seekers I interviewed reported attending such events. A recent study of high-tech job seekers in the Dallas area reports dozens of weekly networking events in that city alone (Lane, 2011). By contrast, there are no networking events in Israel. In fact, most Israeli job seekers had a hard time understanding my question about networking events because they had never heard of such events. Those who understood the question typically had experience working in both the United States and Israel. For example, Jamie, who had worked in both countries, explained that "in Israel, there's no formal networking" but rather "it's seen as a friend thing and there are no networking events or structures of any sort."

Liat, an Israeli currently looking for work in Boston as a radio producer, never attended a networking event while in Israel but has begun to regularly attend such events when looking for work in the United States. She explained the difference:

> In Israel, I never heard of an event where people go and introduce themselves. Here, you need to go to a place and mingle and actually give your

cards. In Israel, the only time you see a business card is from real estate agents. So networking is totally different. In Israel, networking happens if my friend hears of a job opening and they pass along my résumé. There is no physical meeting. There are no gatherings.

The ubiquity of networking events in the United States reflects an unspoken assumption that newly created ties or weak ties can be helpful in one's job search, while their utter absence in Israel reflects the opposite assumption about the futility of attempting to receive help from any such contacts.

To further explore the different approaches to the use of weak ties in Israel and the United States, it is illuminating to compare the use of online social networks. In both Israel and the United States, online social networks, such as LinkedIn and Facebook, have grown rapidly in recent years. Yet, in contrast to the rapid rise in the use of online social networking among job seekers in the United States to reconnect with old contacts—or to connect with their contacts' contacts—Israeli job seekers generally do not use online social networks as part of their job search.

In the United States, job seekers report how social networking sites have facilitated connecting. As Terry puts it: "It's easier, it's convenient. You can do it 24 hours. It's pretty effortless. It's just simpler." Helen reports:

> I use social networking to connect with people that I've worked with that have gone in different ways. There's a tremendous amount of people that I know of that had gone in different ways. If you've lost touch, you can look them up.

Social networking sites are also used by American job seekers to create new ties through introductions. For example, Paul discovered that "a friend of a former roommate of mine who I know very well" was working at a company where Paul himself was seeking employment. He asked the former roommate for an introduction.[9]

Job seekers in Israel, in contrast, do not turn to online social networking in their search. This is not a reflection of a general lack of use of social media. Israelis, in fact, use online social networks even more extensively than Americans.[10] However, they do not use them for job searching. Israelis who are familiar with how LinkedIn is used in both the United States and Israel were best positioned to explain the difference. As Masha put it: "Israelis use LinkedIn, but not like they do in the

United States. They might use it for more social purposes, like to talk to friends. I think it's more for communication."

Israelis' use of online social networks changes, however, when they look for work in the United States. Yaron, an Israeli currently searching for work in Boston, changed his use of LinkedIn following the practice of American white-collar job seekers:

> Yaron: [As part my search] I created a LinkedIn group of Israelis I met in Boston.
> OS: The use of social media is different here from Israel?
> Yaron: Yes. The group I created here I would not have created in Israel. Here, on LinkedIn I am friends with lots of people that I don't really know. In Israel I knew them all.

Similarly, Ari explained his changed use of social networking sites in the United States:

> I added many people to my LinkedIn here that previously I didn't add. In Israel that was less important to me. In Israel no one relies on LinkedIn in hiring... Here LinkedIn is very important. I now connect with many people that I met in Israel that moved to the United States.

Israelis in Boston explained their change of practices in the use of LinkedIn matter-of-factly. Nirit reported that in Israel social networking tools like LinkedIn "are not highly useful," adding that in Israel, "networking is done in other ways." When I probed about these *other ways*, the most common response was that help is sought from family members and close friends, who often date back to the *army days*, the period of mandatory military service that most Israelis go through in their late teens and early twenties. Ronny explained that he is "getting a lot of emails for networking" but, as he put it, "it's not because I'm on LinkedIn." Rather, his networking emails mostly come from friends from his former army unit.

The patterns in the activation of weak ties and the use of online social networks in job searching discussed earlier are summarized in Table 2.

Explaining the Divergence in the Activation of Weak Ties

What explains this cross-national divergence in job-searching practices? What are the conditions that lead job seekers in the

Table 2. Diverging Practices.

Job seekers	Americans in San Francisco and Boston areas	Israelis in Tel Aviv area	Israelis in Boston area
Activate weak ties	86%	0%	75%
Use online social network in job search	71%	0%	60%

United States to pursue weak and new ties while job seekers in Israel focus only on strong ties? Before exploring and theorizing the role of specific labor market institutions, I consider the plausibility of two alternative explanations rooted in broader contextual factors. First, a salient difference among the sites is their respective sizes. For example, the San Francisco metropolitan area, home to over seven million people, is more than double the size of the Tel Aviv metropolitan area with its approximately three million residents (Bay Area Census, 2012; Central Bureau of Statistics of Israel, 2009, Table 2.16). Does the smaller size of Tel Aviv contribute to the predominance of strong ties simply because job seekers are more likely to know each other? Although size may play some role in job seekers' networking practices, it is not a sufficient explanation. Many of the Israeli job seekers I interviewed had no relevant close ties and nevertheless did not consider the possibility of activating weak ties. In fact, there is no verb *to network* in Hebrew; one's network is one's network. Uri, one of the most energetic Israeli job seekers I met, described his personal network as fixed:

> I don't have good friends or family that can help. Some people don't have appropriate connections. What can you do? People around me are blue-collar and don't know high-level people. That's the situation.

This widely shared view suggests that Israelis refrain from reaching out to weak ties even when they lack helpful strong ties. On the flipside, the practices of white-collar Americans in reaching out to weak ties has been documented in locations smaller than the Tel Aviv metropolitan area—such as the Sacramento metropolitan area in California (Smith, 2001)—suggesting again that the size of the local economy is not sufficient to explain the activation of weak ties.

A second broad contextual difference that may be considered important concerns predominant cultural norms. As previously discussed, one strand of the existing literature focuses on cultural explanations by drawing on Hofstede (2001) and theorizing that cross-national variations arise from deeply internalized norms and assumptions regarding the self in relation to others along the dimension of individualism–collectivism (Ma et al., 2011; Xiao & Tsui, 2007). This argument claims that collectivist cultures correspond to distrust of weak ties who are not members of one's in group (Ma et al., 2011; Xiao & Tsui, 2007). Applied to this study, it may be suggested that Israelis' exclusive reliance on strong ties reflects a collectivist culture.

Hofstede's (2001) view on the relationship between cultures and action has been challenged by sociologists of culture who generally do not view national cultures as coherent and deeply internalized systems but as complex and contradictory; instead of providing particular worldviews, cultures are more often seen as offering a variety of tools that actors use in different contexts (Swidler, 1986). The findings in this article support Swidler's (1986) toolkit understanding of culture. Even if modern Israel can be characterized as having a collectivist culture, which is a matter of debate beyond the scope of this article, the findings reported in the prior section—most pointedly, the data discussed with respect to Israeli job seekers who are currently looking for work in the Boston area—show the implausibility of a deeply internalized cultural explanation for cross-national differences in networking practices. If deeply internalized cultural forces were at play, we would expect Israelis who have recently arrived to the United States to maintain a collectivist orientation and to be reluctant to engage in the American style of networking with weak and new ties. We would also expect that this reluctance would be particularly strong when interacting with other Israelis because such interactions occur in Hebrew and generally follow Israeli cultural norms. Yet, the data described in the prior section show the striking matter-of-factness with which Israelis adapt their networking patterns once in the United States. While in Israel, job seekers turn only to strong ties; once in America, they shift gears and turn to weak ties, both Americans and fellow Israelis. Israelis' rapid adoption of American practices of reaching out to weak ties when looking for work in Boston suggests that networking practices are not reflections of deeply internalized cultural norms.

To account for the different strategic approaches toward the use of weak ties, I begin by considering the explanations offered by Israeli job seekers in Boston who pursue weak ties in the United States but did not in Israel. Mor, an Israeli job seeker in Boston, expressed the reasons behind her different approach to weak ties in the United States in a way that mirrors most other Israelis in Boston:

> The use of contacts in Israel does not look so good. Here it is very accepted. In Israel, it was hard to pick up the phone and ask for a favor. To call someone you do not know well—it just would not happen. But here it is very accepted. In Israel you are asking for a favor. You can't call someone that you don't know and say X referred me and maybe you can help me. Here it feels comfortable. Here it's okay. It's understood that this is done. In Israel, I would not have done it. I would have not felt comfortable.

Mor's explanation of her use of weak ties in the United States does not allude to an increase in trust of people outside her in group, as would be suggested by a cultural argument focusing on individualism versus collectivism. Rather, Mor's explanation of her changed approach to weak ties focuses on the different levels of imposition she experiences herself as making on the would-be referrer and the extent to which she feels uncomfortable at burdening the potential referrer by asking for a *favor*.

Because contacting a tie in Israel is understood as asking for a favor, Audrey—like all other Israeli job seekers I interviewed—relies on only close friends and family. She explains that because contacting a tie is asking for a favor, "for contacts to help you ... requires some kind of mutual obligation, something from the past." In the United States, no such past relationship is necessary because in the United States, as Mor put it, "it's understood that it's done."

What underlies this difference? Why does it feel *comfortable* to call a weak tie in the United States but not in Israel? Mor's description of the differences in using weak ties in each context provides a crucial clue. In the remainder of this article, I will unpack the reasons underlying Mor's observation that while in the United States reaching out to weak ties is "very accepted," in Israel turning to weak ties "does not look so good" and is akin to asking for a favor.

A first step to explaining this difference requires looking at job seekers' perceptions of what it takes to get a job in each country. As prior research has shown (Sharone, 2013a, 2013b), Israeli job seekers consistently discuss getting a job as the process of finding an employer looking for their mix of skills, experiences, and credentials—or, as several put it, their *specs*. As Chanan, an Israeli software engineer, explains: "You have to be able to look good on paper [and] fit the keywords" that the employer is looking for. By contrast, white-collar American job seekers view skills and credentials as only one—and not necessarily the most important—factor in hiring. Jason, an American software engineer, succinctly expressed the widely held American view: "The most important thing is fit, not skill. People want to work with people that they like." These different job seeker understandings of what it takes to find a job—which I will refer to as a focus on specs in Israel and on *chemistry* in the United States—reflect how hiring institutions in each site render different dimensions of the applicant-filtering process more *salient* to each group of job seekers.[11]

The specs-focused understanding of the hiring process in Israel is likely rooted in the practices of the dominant screening institutions that job seekers encounter: private staffing agencies and preemployment testing institutes. In Israel, staffing agencies typically post ads, receive and filter stacks of résumés, and conduct initial screening interviews on behalf of employers (Nadiv, 2005; Ram, 2008). *Every* Israeli job seeker I interviewed reported dealing with staffing agencies in his or her search. Job seekers' most salient experiences with staffing agencies are short screening interviews conducted by screeners who use an unbending checklist of credentials, skills, and experiences specified by the client-employer (Sharone, 2013a). The Israeli hiring process also typically involves a daylong examination at a preemployment testing institute (Fizer, 2003). Ninety-one percent of the Israeli job seekers I interviewed reported taking preemployment tests as part of their job search. A typical day of testing combines an examination of writing, math, and logic skills with exercises that aim to measure an applicant's ability to cooperate, communicate, and lead (Sharone, 2013a).

In contrast, the focus of American job seekers on the role of interpersonal chemistry and fit in the hiring process likely stems from hiring practices that make salient to American job seekers the importance of rapport; most significantly, the interviewing practices of hiring managers. In the United States, after an initial behind-the-scenes filtering of résumés by computer programs and human resources personnel, the most common approach to hiring is a direct interview by the hiring manager. Prior research shows that in interviews conducted by hiring managers in the United States, much weight is given to interpersonal fit and rapport (Finlay & Coverdill, 2002; Huo, Huang, & Napier, 2002; Rivera, 2012). As the remainder of this article will show, job seekers' different *perceptions* of how the hiring process works underlie their diverging practices in activating weak ties.

The Activation of Weak Ties in the American Context

The focus of American white-collar job seekers in the United States on developing and using social ties to act as intermediaries between themselves and potential employers—whether such ties are strong, weak, or brand new—is best understood within the broader institutional context that makes salient to job seekers the importance of interpersonal fit and chemistry to the hiring process. My data reveal, consistent with the existing literature (Ehrenreich, 2005; Lane, 2011; Smith, 2001), that American white-collar job seekers who seek guidance on job searching are repeatedly told in books and workshops that the single most important element of the job search is networking. AmeriSupport's director would often repeat the mantra: "Network, network, network. Eighty percent of new jobs are found through your personal network."

The consistent rationale offered for this advice is that hiring managers prefer to hire through referrals because these provide critical information on intangible qualities of the candidate that are not apparent from a paper résumé. The logic is straightforward. Because fit is understood as the key to getting hired and because the best way to convey fit is through a personal referral, finding referrals becomes the central focus of job searching in the United States.

In the American context, a referrer's willingness and ability to make a referral is not perceived to require a long-standing or close relationship. In fact, referrals are widely presumed to be obtainable from anyone, whether a preexisting *weak* tie in Granovetter's (1974) sense, with whom one has not been in touch with for 20 years, or a newly formed tie made at a professional association meeting. The key to getting a referral is generating a sense of interpersonal connection, which conveys having the right intangible qualities to be a good fit. An AmeriSupport speaker provided job seekers with advice on how to create an immediate connection with someone they had just met at a professional association meeting:

> Start by saying, "Sure is a big crowd."...[Then] you say, "I am Mary. I am an accountant. Well, I was until last week when they laid us all off and sent the jobs to India." Notice what I did here. I was very quickly revealing of my need without appearing desperate. That sets the stage for developing the quick connection. Then say, "Tell me about you." [She might respond:] "Oh, I am a marketing manager." But she says it in a flat tone. Part of how you create deep connection instantly is that you listen

not to the words but to what's beneath the words. [Respond] in a tactful way: "Gee, you sound thrilled with that." She gets a little defensive. "Oh no, it's okay. It's just that with my three kids, my plate is pretty full." Aha. You found out her hot button from her animated voice. This is really important. You make an empathetic statement like: "I can understand. How old are your kids?" You share parallel experiences. You show empathy. When you sense she is feeling somewhat connected to you, make the ask: "As I mentioned, I am looking for my next accounting job. Might you know somebody I can speak with?" If she knows someone, she is fairly likely to tell you even though you just met.[12]

Job seeker support organizations like AmeriSupport devote a great deal of time to training job seekers to hone their *elevator speech* or the *30-second me* (Lane, 2011; Smith, 2001) to help their members create an interpersonal connection with networking partners. A good elevator speech is not a recitation of dry facts about oneself, but a communication that will have emotional resonance. Job seekers are reminded that, just like car commercials, the emphasis of the elevator speech should not be on the details of the engine specs. To connect requires that job seekers focus not mainly on the content of their words but rather on their tone and body language.

New members of AmeriSupport devote several hours in their initial week to developing an elevator speech. The most common mistake new members make is to create an elevator speech that sounds formulaic. Career coaches, who come to AmeriSupport as guest speakers, therefore focus their feedback on urging job seekers to be more connecting in their elevator speeches. As one guest speaker told the job seeker group:

> To be honest, when I listened to the people who introduced themselves here, I was unhappy. They were stiff and sterile. It's all about chemistry and connection. If you look like a professional job seeker, it looks like you have been looking for a job too long. Don't do the script with your name stated twice. Stop the scripting. You will look like a loser. Speak your truth informally. Make a connection.

The same focus on creating an interpersonal connection when networking in person also pervades advice on creating one's online social networking profile. One AmeriSupport guest speaker explained: "The way to generate chemistry is to tell me about yourself. Tell me your brand, the kind of person you are. It's a way to decide if they like you."

Most American job seekers dislike networking precisely because the goal of networking—to create interpersonal connection—generates a vulnerability to personalized rejection if the networking effort is not successful (Sharone, 2013b). Online social networking platforms such as LinkedIn are appealing because they promise a way to connect that is much less personally vulnerable. Ultimately, however, most job seekers find that online social networks may be a useful way to identify potential contacts but do not produce enough of a personal connection to generate a referral and therefore cannot substitute for face-to-face meetings. As James, a job seeker, explains:

> [Online social networking] can be useful... But if you allow it to be the sole form of networking—trolling LinkedIn all day has somewhat limited value. It doesn't give you the ability to establish meaningful relationships. People won't recommend you to a hiring manager simply because you met on LinkedIn. But if you reach out to them, have coffee with them, establish a good rapport, have similarities, that's when the person says, "You know what? A certain division within our firm has space. Let me get you in touch with a hiring manager."

The need to create a connection when reaching out to weak ties is the main limitation job seekers in the United States perceive in using social networking sites as a job search tool. As Paul explains: "LinkedIn is easy," but if one only connects online, then one's contacts "don't know anything about you in terms of what your personality is like. Would you be a good fit for the company?" Paul therefore, like James, believes the need to use social networking platforms like LinkedIn to make the initial contact and then set up in-person meetings for coffee or lunch.

Because the end goal of networking in the American context is to produce a connection that will lead to a referral, networking is also one of the most difficult aspects of American white-collar job searching. Connection requires self-revealing communication that will generate a sense of meaningful commonalities. Job seekers typically feel uncomfortable about revealing themselves to strangers. Marsha, for example, reported great hesitation about networking because "I will mess up and people will think I am a dork." It is important to note how the hesitation expressed by Marsha, which stems from the fear of rejection, is different from the previously described Israeli hesitation to contact weak ties, which is rooted in a fear of imposing or of asking a favor of the referrer. The different subjective sources of job seeker discomfort with networking—a fear of rejection in the United States and a fear of imposing on others in Israel—are revealing of different underlying assumptions about the workings of the labor market in each site.

Compared with Israeli job seekers, American job seekers are less concerned about imposing on others. The issue of imposing on others is less salient in the American white-collar context because networking is understood as potentially beneficial not only to the job seeker but also to the referrer and employer. This is because (a) American employers are presumed to be concerned about fit, (b) referrals are widely recognized as the most direct and efficient way to convey fit, and therefore (c) potential referrers are presumed to welcome the opportunity to facilitate a successful match. This understanding, which is at the core of Granovetter's (1974/1995) theory of the motivation of weak ties to make referrals, allows the practice of networking with weak ties to be the center of job searching in the United States. The shared presumption is that employers value information about "the fit between individual personalities and the 'personality' of the job and organization" (Granovetter, 1974/1995, p. 132), and therefore if a referral works out, the referrer is likely to "find their reputation enhanced" (p. 55). As previously discussed, this presumption fits well with dominant understandings of hiring in the American white-collar labor market. As the following section will show, in the Israeli context this presumption is not shared, making the use of weak ties an implausible job search strategy.

The Activation of Weak Ties in the Israeli Context

Although the use of ties to find a job is understood as an efficient way to convey one's fit in the American context, in the Israeli context it is understood as a deviation from the ordinary channels for conveying one's skills and credentials. As Mor, who was earlier quoted, put it: "The use of contacts in Israel does not look so good."

The use of ties *does not look so good* in Israel because it is in tension with the dominant understanding that hiring is about objective skills, an understanding that is discursively and practically reinforced by the dominant Israel hiring institutions—staffing agencies and testing centers. The Israeli objective-skills discourse is, in fact, typically articulated and clarified through a contrast to what many Israeli job seekers call the *old* paradigm of using contacts to *pull strings*. In the Israeli context, the use of contacts is typically depicted as a throwback to an old and corrupt premodern era. The Hebrew word for activating ties, *protectzia*, comes from the Russian word meaning protection—that is, protection from the rigid hiring system and its bureaucratic institutions. It implies getting something because of influence exerted on your behalf as a special favor. Protectzia and the activation of ties—because they introduce emphatically personal and nonprofessional factors—are thus diametrically opposed to an understanding of the hiring process as depending on objective factors. Put simply, the use of personal contacts is understood as a corruption of the system through favoritism.

While the use of protectzia is depicted as a deviation from the proper and legitimate path to employment, most job seekers in Israel nonetheless report that personal connections play an important role in their search. Israeli job seekers who have difficulty finding work often come to see the practices of the staffing agencies and testing institutes as arbitrary. Feeling shut out of the labor market by rigid checklists and tests, they come to see the use of protectzia as a necessary evil. In contrast to the American job search support organizations, where networking is central and much time is spent honing elevator speeches, job seekers at IsraSupport remind each other that, in addition to engaging in their standard job-searching activities, they should also use contacts. As Avi, a job seeker, put it, "remember to also use the 'grandma' method, the old way of using contacts and protectzia."

It is widely assumed that a referral from a close personal connection inside a company can allow one to bypass staffing agency screenings and preemployment testing. Nonetheless, when job seekers talk of using contacts, it is done with either a wink, indicating satisfaction with beating the system, or a sigh, suggesting humiliation at being compelled to engage in ethically questionable practices.

The ironic distance job seekers exhibit about using social ties to find a job is revealed by the tongue-in-cheek way some Israelis refer to using close ties as *taking vitamin P*, with the *P* standing for protectzia. As Michal put it, "you take vitamin P when you look for work where your cousin works or in the business your old army commander now runs."

In interviewing Israeli job seekers, I asked whether they had considered using the networking strategies of American job seekers—turning to weak ties or attempting to develop new ties by approaching strangers at professional conferences. This question typically generated confusion. Audrey, a young lawyer, replied to my question with one of her own: "Why would they refer me if they don't really know me? I would not refer someone that I don't know." As previously discussed, Audrey and other Israeli job seekers in Israel turn to only close friends and family in their job search because they understand the use of ties as imposing a significant burden on the intermediary. Given this assumed burden, it follows that contacts will help only when some mutual obligation from the past exists. The assumed burden on the would-be referrer is rooted in the dominant Israeli understanding of the hiring process in which a referrer must pull strings on behalf of the job seeker and expend capital within his or her organization. This, in turn, is understood to necessitate the preexistence of an ongoing long-term relationship of reciprocal exchanges. In other words, it requires a strong tie.

This difference is not a reflection of a deeply held cultural belief or of collectivist distrust of the *out group*, as shown by the ready use of weak ties by Israelis looking for work in Boston. Rather, it is grounded in the understanding of how hiring works in a particular context. Don, the Israeli job seeker in Boston who was earlier quoted saying that when he last searched for a job in Israel, he had an acquaintance in the company where he interviewed but did not contact him—while if the interview had been in the United States, he would have contacted this acquaintance—explained that the difference in his use of contacts in each place stems from his understanding that in the United States, developing a personal relationship and rapport with the hiring manager is important. Don elaborated: "Here [in the United States], networking is very important. You have to work much on personal contacts, to get to know the hiring managers [through such contacts], and to present yourself to them." Whereas in Israel, contacts are used to obtain favors, in the United States, as Don explains, contacts are used to facilitate a more personal connection.

The different assumptions underlying the use of contacts—most importantly, the different presumptions regarding contacts' potential motivations—are even reflected in the mistakes that Israelis in Boston make when attempting to activate weak ties. Their rapid adoption of the American practice of reaching out to weak ties does not automatically come with an adoption of the American *way* of reaching out. Liraz, an Israeli looking for work in Boston, was surprised to learn that when reaching out to weak ties, it is helpful to meet in person for coffee and not simply make a request for a referral by phone. He explained that in Israel, "there is no need to meet for coffee. Just pick up the phone and ask. This is a favor." Hila, the Israeli career coach who helps Israelis looking for work in Boston, explains that while Israelis rapidly adjust to the practice of contacting weak ties, they are much slower to learn the social protocols involved in making such a contact. The difficult cultural adjustment is not in reaching out to weak ties but in knowing how to reach out. As Hila puts it: "Israelis just call. There is no tango.

No foreplay. No building a relationship." The absence of *tango* stems from the adoption of American weak-tie activation practices without an understanding of the different way the American white-collar labor market actually works; most crucially, the motivations of referrers in the American context where referrals are not understood as requests for a favor but as a match that may benefit all parties involved.

Discussion: Labor-Market Institutions and the Activation of Weak Ties

In-depth interviews with job seekers and participant observations at job search support organizations reveal a sharp cross-national divergence in job seekers' activation of weak ties. Job seekers in Israel exclusively focus on their strong ties, whereas job seekers in the United States, whether culturally American or Israeli, put weak ties at the center of their job search strategies. How do we explain this divergence?

This article suggests that these different strategies are rooted in different job seeker understandings of the workings of the labor market, which arise in different institutional contexts. In the American white-collar institutional context, where employers are perceived to value information about fit and intangible personal characteristics and where the use of ties is understood as an efficient way to convey fit, job seekers reaching out to weak ties perceive a potential outcome that benefits not only themselves but also the weak tie and his or her employer. In the Israeli institutional context, using personal connections is understood as a corruption of the job search process. Because the referring intermediary is asking the employer to make an exception to the usual hiring process based on objective skills, the hiring of a referral involves a chain of linked favors.

First, by deviating from its bureaucratized filtering-and-testing hiring process, an Israeli employer is granting a favor to the referring employee. (In the same way, a study of British and German firms found that hiring an employee's referral is treated as the provision of a "fringe benefit" to the referrer; Wood, 1985, p. 111.) This is a favor because the deviation from the standard hiring process exacts a cost on the organization by setting an undesirable precedent and a cost on the hiring manager for violating organizational norms and expectations.

Second, because the hiring of a referral is a favor granted by the employer to the referrer, the referral itself is also a favor from the referrer to the referred job seeker. In this case, the favor is the expenditure of whatever goodwill or political capital is necessary for the referrer to make the referral to the employer.[13] For this reason, Israeli job seekers reasonably presume the need for a strong tie with a past history of mutual and reciprocal obligations. In short, because in the Israeli context the use of connections to exert influence imposes a high cost on the intermediary, American-style networking focused on weak or new ties is perceived as impractical and is therefore not practiced in Israel.

The Israeli chain-of-favors dynamic stands in sharp contrast to the American context, where the person acting as the intermediary is not asking the hiring manager for a favor but assumes the role of a helpful matchmaker bringing together two suitable parties for their mutual benefit. Since networking is understood in the American context as a legitimate and efficient practice, there is far less cost and risk to acting as an intermediary. In the context of a hiring system with a significant focus on interpersonal fit or chemistry, making a referral entails only convincing the decision maker that one has relevant information about a potential candidate's fit. While a chain of favors is required for referrals to be effective in the Israeli context, an American hiring manager is often grateful to the intermediary for providing relevant information. In fact, as Granovetter (1974/1995) recognized, in the American context weak ties are motivated to make referrals precisely because of this potential benefit of enhancing their reputations as contributors to the employer. Granovetter's understanding of the reputational benefits to successful referrers is based on the widely shared premise in the American white-collar labor market that referrers are making a valuable contribution to employers by providing important but difficult-to-obtain information about candidates' intangible qualities and likely fit with the employer. As Granovetter (1974/1995) explains, unlike quantifiable information, "some kinds of data cannot be numerically reproduced...the best examples involve the fit between individual personalities and the 'personality' of the job and organization" (p. 132).

Making a referral is never risk-free. In both the American and the Israeli contexts, the referrer bears some reputational risk in making the referral (Marin, 2012; Smith, 2005, 2007) because referring an employee who does not perform can damage the reputation of the referrer. But the risks of a mismatch, which are common to both countries, need to be distinguished from the risks and costs of proposing a match in the first place. In Israel—but not in the United States—even if the match is ultimately successful, the referrer incurs costs in making the referral because he or she is necessarily asking the hiring manager to deviate from the standard hiring process.

Conclusion

Returning to the question posed at the start of this article regarding whether and how institutional contexts encourage or inhibit the activation of weak ties, this article shows that different institutional contexts generate different job seeker understandings of the hiring process, which in turn can make the activation of weak ties the central focus of the job search, as is the case in the United States, or can make such ties largely irrelevant, as is the case in Israel.

There are significant implications to the finding of this article that the activation of weak ties is not an inherent feature of advanced white-collar labor markets but instead a contingent outcome of particular institutional arrangements. For example, a growing literature reveals that when hiring outcomes depend on job seekers possessing broad networks, the results exacerbate gender and racial inequalities (for a recent review, see Trimble & Kmec, 2011). If the strength of weak ties is not inherent to advanced market economies but a feature of the American white-collar institutional context, then these unequal outcomes can be addressed at their root through public policies and institutional reforms. Regarding future research, the findings of this article suggest that to understand the use of different forms of social capital requires careful examination of how institutional contexts shape actors' understandings of the meaning of activating a referral. As Granovetter (1974/1995) has taught a generation of economic sociologists, actors are embedded in institutions and "job searching behavior is more than a rational economic process—it is heavily embedded in other social processes that closely constrain and determine its course and results" (p. 39). Further research is needed to explore the range of institutional contexts that generate variations in the activation of weak ties and the social processes that underlie such variations.

Acknowledgments

I would like to thank Roberto Fernandez, Mathew Bidwell, Peter Cappelli, Emilio Castilla, Ana Villalobos, Sandra Smith, and the anonymous reviewers of *Work and Occupations* for their feedback and support on this manuscript.

Declaration of Conflicting Interests

The author(s) declared no potential conflicts of interest with respect to the research, authorship, and/or publication of this article.

Funding

The author(s) received no financial support for the research, authorship, and/or publication of this article.

Notes

1. After California, Israel has the second highest concentration of high-tech companies in the world (Rosenberg & Vainunska, 2007).
2. See http://techcrunch.com/2012/11/20/startup-genome-ranks-the-worlds-top-startup-ecosystems-silicon-valley-tel-aviv-l-a-lead-the-way/.
3. It may be correctly pointed out that Israel and the United States have very different political-economic histories (Sharone, 2013b). Although the role of the state and of unions in the labor market has declined in both countries, this decline was more precipitous in Israel (Farber, 2008; Kalleberg, 2011; Osterman, 1999; Ram, 2008; Samuel & Harpaz, 2004). These differences notwithstanding, with respect to white-collar workers—the focus of this article—the broad historical change from security to precariousness is shared. In both cases, white-collar workers enjoyed job security during the post-World War II era and until the 1970s—though the source of this security was different—and since that time, in both sites, workers have been adjusting to the loss of this security (Ram, 2008; Smith, 2001).
4. American job seekers in the San Francisco and Boston areas and Israeli job seekers in the Tel Aviv area were interviewed in 2005–2006 and 2010–2012. Israelis searching for work in the Boston area were interviewed in 2011–2012.
5. I chose to focus on unemployed job seekers due to their typically conscious and active consideration of different job search and networking strategies. This focus, however, is also a limitation of this study because unemployed job seekers may engage in networking practices that are different from those of employed job seekers.
6.

Occupation	San Francisco	Tel Aviv	Boston
Technical	35%	39%	40%
Managerial	28%	24%	10%
Professional	13%	10%	15%
Other white collar	24%	27%	35%

7. The self-selection mechanism for participating in these support groups appears similar across Israel and the United States. Comparing my data for job seekers who belong to such groups and those who do not suggests that the groups attract job seekers who typically differ from the general population in two ways: First, at the point of joining, they are less likely to have promising job leads through their preexisting networks and therefore feel more anxious about their employment prospects. Second, from among job seekers who are facing difficulties, those who join support groups are

more motivated than other job seekers to seek advice about job-search strategies. Thus, both the most well-connected and the most discouraged job seekers whom I interviewed did not attend these groups. On the central question explored in this article—the activation of weak ties—there was no significant difference in the responses of job seekers who were members of support groups and those who were not.
8. Granovetter (1974/1995) claimed:

> It is all but impossible to arrange one's interactions in such an artificial way. The idea that one was being "cultivated" as a potential labor market contact would offer strong motivation to curtail the relationship. Such proposals can arise only through failure to recognize the extent to which labor-market behavior is embedded in other economic and social activity. (p. 133)

9. Granovetter (1974/1995) amazingly anticipated this technology in the 1970s, long before it arose. Based on his findings that weak ties are effective even if they work through "chains" of one or two intermediaries, in the policy section of his book Granovetter considers the possibility of creating a database that would help job seekers figure out whom they know that knows the targeted employer (p. 137).
10. Based on a combination of average daily visitors and page views, LinkedIn is the ninth most popular site in Israel, while it ranks eleventh in the United States. See http://www.appappeal.com/maps/linkedin/.
11. It is beyond the scope of this article to determine the accuracy of job seekers' perceptions about the hiring process in each site. For my argument, it is sufficient to observe that institutions make *salient* different filters used in the hiring process and that job seekers' understandings of the hiring process, as formed in this institutional context, underlies the observed pattern in weak-tie activation.
12. Sandra Smith's (2012) research focusing on the behaviors of referrers suggests that, in the United States, referrals are indeed made to employers based on a very limited *thin slice* instantaneous assessment.
13. It is important to note that even if some Israeli employers take a more American approach to referrals and view these as helpful, this would not likely alter job seekers' willingness to approach weak ties; they are unlikely to know which of their weak ties work for such an exceptional employer.

References

Aral, S., & Alstyne, M. V. (2011). The diversity-bandwidth trade-off. *American Journal of Sociology*, *109*(6), 319–364.

Bay Area Census. (2012). *San Francisco Bay area*. Retrieved from http://www.bayareacensus.ca.gov/bayarea.htm

Bian, Y. (1997). Indirect ties, network bridges, and job search in China. *American Sociological Review*, *62*, 366–385.

Burt, R. (1977). The contingent value of social capital. *Administrative Science Quarterly*, *42*(2), 339–365.

Central Bureau of Statistics of Israel. (2009). *Localities, population and density per sq. km. by metropolitan area and selected localities 2008* (Statistical Abstract of Israel). Retrieved from http://www.cbs.gov.il/reader/shnaton/shnatone_new.htm?CYear=2009&Vol=60&CSubject=2

De Graaf, N., & Flap, H. D. (1988). "With a Little Help from My Friends": Social resources as an explanation of occupational status and income in West Germany, the Netherlands, and the United States. *Social Forces*, *67*(2), 452–472.

De Schweinetz, D. (1932). *How workers find jobs*. Philadelphia: University of Pennsylvania Press.

Ehrenreich, B. (2005). *Bait and switch: The (futile) pursuit of the American dream*. New York, NY: Metropolitan Books.

Farber, H. S. (2008). "Short(er) Shrift": The decline in worker-firm attachment in the United States. In K. S. Newman (Ed.), *Laid off, laid low: Political and economic consequences of employment insecurity* (pp. 10–37). New York, NY: Columbia University Press.

Finlay, W., & Coverdill, J. (2002). *Headhunters: Matchmaking in the labor market*. Ithaca, NY: ILR Press.

Fizer, C. (2003). *Preparing for employment tests* (in Hebrew). Tel Aviv, Israel: Center for Occupational Advice and Science.

Gal, J. (2005). The rise and fall of unemployment insurance in Israel. *International Social Security Review*, *58*, 107–116.

Gangl, M. (2004). Welfare states and the scar effects of unemployment: A comparative analysis of the United States and West Germany. *American Journal of Sociology*, *109*(6), 1319–1364.

Gerber, T., & Mayorova, O. (2010). Getting personal: Networks, institutions, and stratification in the Russian labor market, 1985–2001. *American Journal of Sociology*, *116*(3), 855–904.

Granovetter, M. (1973). The strength of weak ties. *American Journal of Sociology*, *78*, 1360–1380.

Granovetter, M. (1974/1995). *Getting a job: A study of contacts and careers* (2nd ed.). Cambridge, MA: Harvard University Press.

Hipp, L. (2011, May). *Contracts, confidence, and continuous employment: The relationship between labor market policies and perceived job security* (PhD Dissertation). Cornell University, Ithaca, NY.

Hofstede, G. (2001). *Cultures' consequences: Comparing values, behaviors, institutions and organizations across nations*. Thousand Oaks, CA: Sage.

Huo, P., Huang, H. J., & Napier, N. (2002). Divergence or convergence: A cross-national comparison of personnel selection practices. *Human Resource Management, 41*(1), 31–44.

Kalleberg, L. A. (2009). Precarious work, insecure workers: Employment relations in transition. 2008 ASA presidential address. *American Sociological Review, 74*(1), 1–22.

Kalleberg, L. A. (2011). *Good jobs, bad jobs: The rise of polarized and precarious employment systems in the United States, 1970s–2000s*. New York, NY: Russell Sage Foundation.

Korpi, T. (2001). Good friends in bad times? Social networks and job search among the unemployed in Sweden. *Acta Sociologica, 44*(2), 157–170.

Lane, C. M. (2011). *A company of one: Insecurity, independence, and the new world of white-collar unemployment*. Ithaca, NY: Cornell University Press.

Ma, R., Huang, T., & Shenkar, O. (2011). Social networks and opportunity recognition: A cultural comparison between Taiwan and the United States. *Strategic Management Journal, 32*, 1183–1205.

Marin, A. (2012). Don't mention it: Why people don't share job information, when they do, and why it matters. *Social Networks, 34*(2012), 181–192.

Marsden, P., & Hulbert, J. (1988). Social resources and mobility outcomes: A replication and extension. *Social Forces, 66*(4), 1038–1059.

Mishel, L., Bernstein, J., & Allegretto, S. (2007). *The state of working America 2006/2007*. Ithaca, NY: Cornell University Press.

Nadiv, R. (2005). *Licensed manpower companies in Israel*. Jerusalem, Israel: Report of the Israeli Parliament Research and Information Center.

Obukhova, E. (2012). Motivation vs. relevance: Using strong ties to find a job in urban China. *Social Science Research, 41*(3), 470–480.

Osterman, P. (1999). *Securing prosperity: The American labor market, how it has changed and what we can do about it*. Princeton, NJ: Princeton University Press.

Ram, U. (2008). *The globalization of Israel*. New York, NY: Routledge.

Reagans, R., & McEvily, B. (2003). Network structure and knowledge transfer: The effects of cohesion and range. *Administrative Science Quarterly, 48*(2), 240–267.

Rivera, L. A. (2012). Hiring as cultural matching: The case of elite professional service firms. *American Sociological Review, 77*(6), 999–1022.

Rosenberg, Y. & Vainunska, K. (2007). *Israel: The new silicon valley* (Unpublished master's thesis). Tel Aviv University Faculty of Management.

Samuel, Y., & Harpaz, I. (2004). *Work and organizations in Israel: Studies of Israeli society* (Vol. 11). London, England: Transaction.

Sharone, O. (2013a). Why unemployed Americans blame themselves while Israelis blame the system. *Social Forces, 91*(4), 1429–1450.

Sharone, O. (2013b). *Flawed system/flawed self: Job searching and unemployment experiences*. Chicago, IL: University of Chicago Press.

Smith, S. (2005). "Don't put my name on it": Social capital activation and job finding assistance among the black urban poor. *American Journal of Sociology, 111*(1), 1–57.

Smith, S. (2007). *Lone pursuit: Distrust and defensive individualism among the black poor.* New York, NY: Russell Sage Foundation.

Smith, S. (2012, October). *Why weak ties help and strong ties don't: Reconsidering why tie strength matters.* Paper presented at Institute for Work and Employment Research workshop, MIT Sloan, Cambridge, MA.

Smith, V. (2001). *Crossing the great divide: Worker risk and opportunity in the new economy.* Ithaca, NY: Cornell University Press.

Swidler, A. (1986). Culture in action: Symbols and strategies. *American Sociological Review, 51*, 273–286.

Trimble, L., & Kmec, J. (2011). The role of social networks in getting a job. *Social Compass, 5*(2), 165–178.

Wood, S. (1985). Recruitment systems and the recession. *British Journal of Industrial Relations, 23*(3), 103–120.

Xiao, Z., & Tsui, A. (2007). When brokers may not work: The cultural contingency of social capital in Chinese high-tech firms. *Administrative Science Quarterly, 52*(1), 1–31.

Yakubovic, Y. (2005). Weak ties, information, and influence: How workers find jobs in a local Russian labor market. *American Sociological Review, 70*(3), 408–421.

Author Biography

Ofer Sharone is an assistant professor of Work and Organization Studies at MIT Sloan. Sharone's research focuses on career transitions and unemployment. His recently published book, *Flawed System/ Flawed Self: Job Searching and Unemployment Experiences* (University of Chicago Press), compares the experiences of job seekers in Israel and the United States.

Strong Ties, Weak Ties, or No Ties: What Helped Sociology Majors Find Career-Level Jobs?

ROBERTA SPALTER-ROTH[†]
NICOLE VAN VOOREN[†]
MICHAEL KISIELEWSKI[†]
MARY S. SENTER[††]

JULY 2013

DEPARTMENT OF RESEARCH ON THE DISCIPLINE AND PROFESSION • AMERICAN SOCIOLOGICAL ASSOCIATION

INTRODUCTION

Many sociology departments in schools of liberal arts and sciences are concerned about losing out in the competition for undergraduate majors, even though a study of senior sociology majors who graduated in 2012 by the American Sociological Association (ASA) showed that students were excited by sociological concepts and were very satisfied with the major (Senter et al. 2012; Spalter-Roth et al. 2012). Given that today's college students are entering a job market with the highest unemployment in a generation, and are saddled with increasing debt (Baum and McPherson 2010), it is reasonable that students and their parents have been concerned about the prospects of students obtaining professional or "career" jobs. The unemployment rate for sociology baccalaureates was 9.9%, almost twice that of those with degrees in nursing, but slightly less than for graduates in political science and about 3% less than for anthropology graduates (Carnevale and Cheah 2013). A lack of understanding about how sociology majors search for and secure career jobs, and the kind of social capital that helps them in this process, can result in sociology departments losing majors to more vocationally-oriented programs in fields such as health (Brint 2002, 2005, 2010).[1]

This research brief is based on responses from the first and second waves of ASA's longitudinal survey of senior majors from the class of 2012. In it we asked what type of jobs these majors obtained about seven months after graduation; which of these jobs survey respondents considered to be career-type jobs as opposed to jobs that were not on a career-track; what social capital in the form of social ties and resources students used in their job search; and how effective these sources were for obtaining career-level positions. Knowledge about what kinds of social capital are important for a successful career search will be increasingly important as employment outcomes for graduates become one of the five metrics used for accountability to policymakers (Gardner 2013).

DEFINITIONS

Social capital has been defined as "the various resources embedded in networks that can be accessed by social actors," especially in the job search process (Bourdeiu 1986; Lin 2001 cited in Martin 2009). Although this term is widely used in sociology, Martin (2009) claims that "little research has examined social capital at the post-secondary level"

[†] Department of Research on the Discipline and Profession, American Sociological Association.

[††] Central Michigan University.

[1] Sociology faculty might well be concerned about the professional job prospects of their majors because they care about the success of their students and believe that sociologically-trained graduates will contribute to societal well-being.

(p. 187). When research is available, it suggests that resources gained through various networks are beneficial for college students (Pascarella and Terenzini 2005). In contrast, other researchers have suggested that social capital is not directly related to using contacts for job search or the prestige of the job obtained (Mouw 2003). However, in his study, Montgomery (1994) finds that weak ties are related to higher wages and higher rates of employment because these ties provide new information by taking people out of their immediate networks. Other studies posit that strong ties are not useful for most job searches, but that classifications such as strong and weak are too simplistic (Granovetter 1995; Grannis 2010; Rosenbaum, DeLuca, Miller, and Roy 1999; McDonald and Elder 2006; Wegener 1991).

According to the accepted literature in the field, more than half of job seekers find their positions through social capital in the form of personal ties (Fernandez, Castilla, and Moore 2000; Marsden and Gorman 2001; Reskin 1998; McGuire 2007; Neckerman and Fernandez 2003; Neckerman and Kirshenman 1991). However, it is not clear whether these ties are weak, characterized as non-frequent and transitory relations (Montgomery 1994), or strong, characterized by emotional intimacy, intensity, and trust (Krackhardt 1992). A third type of tie can be referred to as absent ties, for which we have no information, or impersonal ties, in which there is no face-to-face interaction.

Students can draw social capital in the form of social relationships, contacts, and resources from a variety of sources, both personal—either strong or weak—and impersonal, in their search for post-baccalaureate employment and graduate education. Parents can provide these sources of social capital for job searches, but previous research suggests that help from this strong tie is more likely when parents (especially white males) themselves have high amounts of human capital including graduate degrees (O'Reagan and Quigley 1993). Relationships with faculty members can result in vocational preparation as well as intellectual development (Kuh and Hu 2001), and extracurricular activities can result in career-relevant skills (Pascarella and Terezini 2005). Frequent interactions with peers can result in commitment to college programs and knowledge and skill acquisition, although not related to GPA (Martin 2009).

This brief examines the social capital that sociology majors called on for help in their post-graduation job search and the results of the search. We asked whether senior majors had the social capital in the form of relationships and the resources that they needed to guide them in obtaining professional or career-level jobs, and whether these relations were weak or strong. In contrast to strong or weak ties, did they use impersonal resources that may involve no ties? Specifically, which relations and resources led to career-type occupations and which did not? We looked into whether or not the type of institution of higher education attended affected the type of ties and outcomes. We examined further how the demographic characteristics of graduates affected the type of social capital they called on. All students may have used a combination of personal and impersonal relations and resources in their job search, although Martin (2009) notes that black students have fewer relations than whites because individuals tend to pick or have access to people like themselves as contacts (McPherson, Smith-Lovin, and Cook 2001 cited in Mouw 2003).

In addition, using optimal scaling analysis, we examined whether the different types of ties and resources formed clusters that reflected strong, weak, and impersonal or no ties. Finally, using logistic regression analysis, we determined which measures were significantly related to obtaining career-level positions when controlling for other characteristics such as gender, race and ethnicity, type of institution of higher education, and fathers' education.

The survey design for this study is described in Appendix I.

FINDINGS

The first set of findings presented is descriptive, based on frequencies and cross-tabulations, while the second set of findings is based on cluster or multivariate analysis.

CAREER JOBS VS. NON-CAREER JOBS

In the tough job market for 2012 sociology graduates, 56% of respondents were working at a paid job or paid internship (only), and another 20% were both working and enrolled at a college or university. The remaining 24% were attending graduate school but not working. The analysis of job search included three groups: those who were working (only), those who reported that their primary activity was

Table 1. Number in Job Category and Percentage of Job Holders Who Perceive They Have a Career Job: 2012.

	N in Job Category	Percent
Management-related	28	96.4
Social Science Researcher	33	81.8
IT, PR, Other	60	81.7
Social Services/Counselors	170	71.8
Teachers	84	67.9
Clerical/Administrative assistant	101	50.5
Sales/Marketing	105	39.0
Service Occupations	82	17.1
Other	56	60.7

Source: American Sociological Association. *Social Capital, Organizational Capital, and the Job Market for New Sociology Graduates Survey*, 2012.

working at a job although they attended graduate school as well, or those who had searched for a position but did not find one. The analysis of careers included the first two groups (with a total of 715 respondents). More than half (58.9%) of these graduates who said that work was their primary activity obtained what they perceived to be career-level jobs, with slightly more than 40% failing to do so.[2] Table 1 ranks the percentage of respondents who agree that the occupational category in which they landed in 2013 was a career-type position. The table shows that close to 100% (96.4%) of those who found management-related positions think that these jobs would lead to a professional career. Yet, the table also shows that this was the smallest category of occupational participation. The second-smallest occupational category was social science researchers, although 81.8% agree that they held a career-type job. Previous research suggests that graduates of sociology terminal master's programs were much more likely to hold research-positions than baccalaureates (30.4% compared to 4.6%, respectively), undoubtedly because they had more methods and statistics training (Van Vooren and Spalter-Roth 2011). The third-smallest job category, information technology (IT) and public relations (PR), had a similar percentage of post-baccalaureate sociologists agreeing that they were in a career or professional position. The largest job category in 2013, seven months after graduation, was social service workers and counselors with 71.8% agreeing that these were career-level jobs. In contrast, the next two largest jobs categories, clerical/administrative assistant and sales and marketing, had significantly lower percentages of respondents who agreed that these were career-type jobs. Only about half (50.5%) of clerical workers and 39.0% of sales workers agreed that these were career-level positions. Service occupations were the least likely to be thought of as career positions, with only 17.1% of respondents agreeing that they were.

CHANGES SINCE 2005

The pre-Recession class of 2005 fared somewhat better that the post-Recession class of 2012 in terms of their ability to obtain career-level positions. Table 2 shows that, in general, the percent of sociology baccalaureates in professional level jobs decreased somewhat pre- and post-Recession, while the percentage of graduates in non-professional jobs increased somewhat. Between these years, there was a dip in social service, IT and PR positions, while the percent of graduates in social science research jobs stayed relatively stable. The largest decrease was in management-related positions, thought to be the most career-related of all positions. However, the large drop in management jobs might be an artifact either

Table 2. Job Categories of Sociology Majors, Wave 2, 2005 and 2012: Percentage and Percentage Point Change.

	2012 (N=759)	2005 (N=621)	Percentage point change
Social Services/Counselors	23.7	26.5	-2.1
Sales/Marketing	14.2	10.1	4.1
Clerical/Administrative Assistant	14.0	15.8	-1.8
Teachers	11.9	8.1	3.8
Service Occupations	11.9	8.3	3.6
IT, PR, Other	7.9	10.2	-2.3
Social Science Researcher	4.6	5.7	-1.1
Management-related	3.8	14.4	-10.6
Other	8.0	4.4	3.6

Source: American Sociological Association. *Social Capital, Organizational Capital, and the Job Market for New Sociology Graduates Survey*, 2012; and *What Can I Do with a Bachelor's Degree in Sociology?* (Wave 2), 2007.

---

[2] The determination of whether a job was career-level or not resulted from responses to the question "Is this a career-type job?"

of the timing of the second wave of the survey or of self-coding.[3] Teaching was the only career-oriented position (with 60% of respondents so-labeling it) that increased. In contrast to decreases in career-type positions, jobs that respondents agreed were not on a career track increased somewhat. The two-largest increases were sales and marketing jobs and service positions, while the percentage of graduates in clerical/administrative support positions stayed relatively stable (see Table 2).

STUDENT BACKGROUNDS

We look at two characteristics of students' backgrounds—the type of institution of higher education that they attended and the educational background of their parents.

Relationship between type of institution of higher education and job type. In a previous research brief we found that senior sociology majors were not very satisfied with the career counseling that they received, although those at Master's comprehensive universities were significantly more satisfied (35% were) with career counseling than seniors at either Research and Doctoral universities or Baccalaureate-only colleges (Senter et al. 2012). Yet, when we examined whether or not the former major obtained a career-type occupation, we found no significant differences by type of school (as measured by Carnegie Codes; see Figure 1).

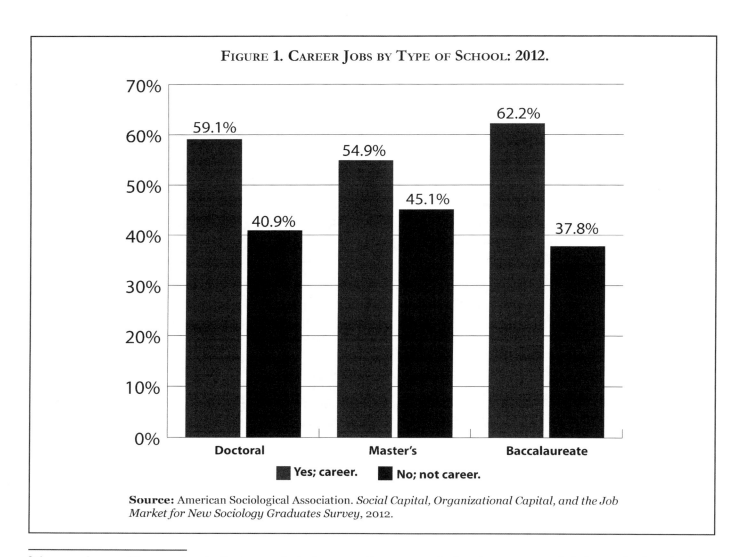

FIGURE 1. CAREER JOBS BY TYPE OF SCHOOL: 2012.

Source: American Sociological Association. *Social Capital, Organizational Capital, and the Job Market for New Sociology Graduates Survey,* 2012.

---

[3]The second wave of the 2005 longitudinal survey took place more than one year (rather than seven months) after graduation. Spalter-Roth and Van Vooren (2010) indicate in previous research that graduates tend to receive promotions into management-type positions as the time since graduation increases.

Relationship between parents' education and job type. In our previous research about this cohort we found that more than half of respondents' mothers did not have a college degree, and almost 30% of these mothers had a high school degree or less. Fathers had similar levels of education to mothers (Senter et al. 2012). The parents of students at Baccalaureate institutions were most likely to be highly educated: more than one-third of these students reported that their parents had at least some graduate or professional schooling. By contrast, only about 15% of students at Master's institutions reported mothers or fathers who were this highly educated. These data suggest that, especially at Master's institutions, respondents were first generation college graduates. Overall, Table 3 does not show a relationship between parent's education and whether their children obtained a career-level position. Although the overall relationship is not significant, 64% of the sociology majors whose fathers who attended graduate school or had a graduate degree were likely to have obtained a career-level job, about 10 percentage points higher than the children of fathers with less education.

Table 3. Parents' Level of Education by Professional Job (Percent in a Professional Job).

	% In a Professional Job: Father's Education	% In a Professional Job: Mother's Education
High school graduate or less	54.4	54.0
Associate/nursing degree	55.8	60.7
College graduate	56.1	58.0
Graduate degree	64.3	56.8

Source: American Sociological Association. *Social Capital, Organizational Capital, and the Job Market for New Sociology Graduates Survey*, 2012.

JOB SEARCH

Use and Effectiveness of Job Resources

During the period that Granovetter (1973, 1983, 1995) and others were writing as to whether strong or weak ties resulted in effective job search, the availability of online search techniques was nonexistent or extremely limited. In contrast, the class of 2012 was most likely to use online job search sites such as Monster.com, Idealist.org, or Craigslist as a job-search strategy. More than half (56.0%) of former sociology majors reported using these online sources, and of all the techniques used, online sources were reported to be the most effective during the job search. The next most commonly used source was a classmate, colleague, or friend, closely followed by a family member, with effectiveness scores hovering about 50% (54.2% and 48.4%, respectively—see Table 4).

Unlike online searches that do not include direct social relations, family and friends were viewed as strong ties by study respondents. Two measures were used to quantify the strength of these ties: the responses to the questions, "how close are you to this person?" and "how often do you interact with this person?" The responses ranged from 1-"not at all close" through 5-"very close." Family members received median scores of 5 on both closeness and frequency of interaction—the highest possible score. Friends and classmates received a median closeness score of 4 and an interaction score of 3.

A third social tie was with a faculty member with a median closeness score of 3, neither strong nor weak. However as Table 4 shows, fewer than 12% of students use faculty members for help in their job search and

Table 4. Use and Effectiveness of Job Search Methods.

	% Using	N Using	% Most Effective	N Most Effective
Online methods	56.0	459	68.0	312
Classmate, colleague, friend	35.9	294	54.2	160
Family member	33.0	271	48.4	132
College career services	22.8	187	40.6	76
Newspaper ads	16.3	134	15.7	21
Employment agency	12.8	105	39.0	41
Former employer	12.1	99	55.6	55
Faculty	11.5	94	13.5	13
Job/Internship supervisor	11.5	94	51.1	48
Unsolicited resume	5.1	42	21.4	9
Workshop in sociology department	2.1	17	35.3	6
Capstone	1.2	10	30.0	3

Source: American Sociological Association. *Social Capital, Organizational Capital, and the Job Market for New Sociology Graduates Survey*, 2012.

only 13.5% of those who called on faculty members found their help to be effective in searching for jobs. A similar percent of students called on job or internship supervisors, yet these supervisors were viewed as 3.7 times as effective as were faculty members. Likewise, about the same percent of respondents called upon former employers and in this case they were regarded as more than four times as effective in aiding sociology majors' job search as were faculty members.

About half (53.3%) of respondents used two or three job search methods, and about one-third used four or more methods. There were significant differences in the mean number of searches used by those who found career-level positions and those who did not; however, it was those who used one method who were most likely to secure a career-type job (see Figure 2). Possibly, these respondents were more focused and particular in their job search method rather than using a less mindful "scatter-shot" approach. Alternatively, these graduates may have had considerable difficulty finding a job, tried many sources to secure employment, and settled for a non-professional job.

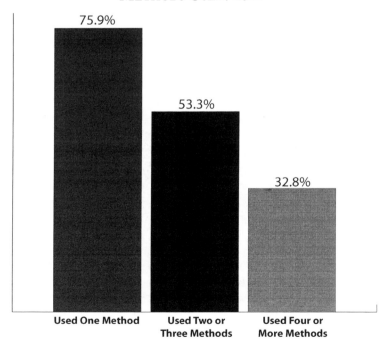

Figure 2. Career Jobs by Number of Search Methods Used: 2012

Source: American Sociological Association. *Social Capital, Organizational Capital, and the Job Market for New Sociology Graduates Survey*, 2012.

Job Search, Ties, and Career-Level Employment

We saw that despite the tough 2012 job market, with an unemployment rate of about 10% for graduating senior sociology majors, more than half of respondents agreed that they had obtained a career-level job about seven months after graduation. In Table 5 we show the type of ties that resulted in career-level positions and those that did not. In general, what would be described as weak ties were more likely than strong ties or no ties to result in career-type jobs.

Weak Ties. Although capstone courses appeared to be the job search method most likely to result in a career-type position (with 87.5% of those who used this method of job search having found a career-type job), fewer than 2% of respondents had used this method (refer back to Table 4). This finding suggested that only a relatively small number of departments offered capstone courses with a career emphasis to graduating seniors; the low N may, therefore, may make this finding unreliable. Other weak ties such as job or internship supervisors, former employers, college career services, and career workshops in the sociology department (here again relatively few students participated in such workshops) resulted in between 57.9% and 70.6% of respondents obtaining career-type jobs (see Table 5). For those respondents who sought advice from faculty members (who were rated in the middle of the closeness scale), 62.0% found career-level positions. Recall, however, that only 11.5% of respondents used this technique and only 13.5% said it was the most effective job search method. The relatively high percent of respondents that believed they found a career-level job using this search technique suggests that faculty, who collect information about careers or encourage students in job search techniques, may be undervalued as a source of job search information.

Strong Ties. Strong ties with classmates and friends or family members/relatives did not appear to be the most effective techniques for finding career-level or professional positions. Almost half of respondents who relied on classmates and friends (48.8%) and more than half (54.8%) of those who used family or relatives as a major job search resource did not find career-level positions. In fact, those respondents who used a family member as a resource were

Table 5. Career Jobs by Job Search Method: 2012 (Percent that Used Each Method).

Method	In a Career Job (%)
Capstone	87.5
Job/Internship supervisor	70.6
Other	70.6
Former employer	67.0
College career services	57.9
Faculty	62.0
Workshop in sociology department	58.8
Unsolicited offer	54.5

Method	Not in a Career Job (%)
Classmate, colleague, friend	48.8
Family member	54.8
Online methods	55.0
Unsolicited resume	59.4
Employment agency	71.8
Newspaper ads	81.0

Source: American Sociological Association. *Social Capital, Organizational Capital, and the Job Market for New Sociology Graduates Survey*, 2012.

most likely to have found work as a sales or a clerical worker. These findings suggest that strong ties are not necessarily the best resources for those sociology majors looking for a professional-level position.

Impersonal or Non-Ties. Impersonal ties or non-ties were the least likely to result in career-type positions and the most likely to result in non-career-level jobs such as service, clerical, and sales work. Of those who used newspaper ads as a search method, fully 81.0% did not find career-level positions, followed by 71.8% of respondents who used employment agencies. As we saw, the largest number of respondents reported using online search methods and thought of them as effective methods for finding jobs even though 55% ended up in positions that they did not consider professional-level. These findings suggest that sociology graduates (as with others in their age cohort) appeared to overestimate the value of the impersonal job search.

GROUPINGS OF JOB SEARCH STRATEGIES

We have seen that some strategies are more likely to result in what these respondents classified as either career-type or non-career-type jobs, but did any of the strategies form scales or group together? In other words, by using factor analysis could we assign the strategies to groups or categories that fit with the types of ties that we have discussed, i.e., weak, strong, or impersonal or non-ties?[4] The answer was yes. The first analysis resulted in six clusters that explained 56% of the variance. Two of the groups included weak ties, two included impersonal ties, and one included personal ties. The first of the two clusters used in job searches that represented weak ties included faculty members and capstone courses. The second cluster of weak ties included former employers and internship advisors. Clusters that represented impersonal or non-ties included newspapers and employment agencies for one, and unsolicited resumes and receiving unsolicited offers in the other. The impersonal strategy that was used most by respondents, online job search, stood alone. A cluster of strong ties included friends and classmates, on the one hand, and family members or relatives on the other. A number of variables did not scale, including job workshops, career services, and other methods. The strategies that did cluster suggested that the notion of strong ties, weak ties, and impersonal ties were valid constructs (see Figure 3).

In order to pursue further the issue of whether job search strategies group together, rather than allowing SPSS to automatically determine the number of factors in each grouping, we selected four clusters to be extracted.[5] The first scale or grouping that emerged included strategies that we have referred to as the use of weak ties, the second included strategies that we have referred to as strong, personal ties, the third includes strategies that are impersonal ties, and a fourth, less strong cluster, that included a weak tie and an impersonal tie (see Figure 4). The "impersonal tie" cluster included newspaper ads, employment agencies, and online searches. (Another impersonal strategy, sending unsolicited resumes, was not part of this or any other cluster.) The second cluster of strategies included what we have labeled as "strong ties,"

---

[4] We used Principal Components Analysis to develop the scales or groupings in order to discover whether the concept of different types of ties appeared to be valid. First, we did an analysis that included two variables in each set and followed with an analysis that included four variables in each set. The results, as discussed above, did not appear to be very different. All of the strategy variables were included in the scaling procedure, but only those that showed significant differences were included in the scales. The rotation method used for this analysis was Varimax with Kaiser Normalization, and the rotation converged in six iterations.

[5] We used the scree plot that is part of the output from the Principle Components Analysis to do the second factor analysis.

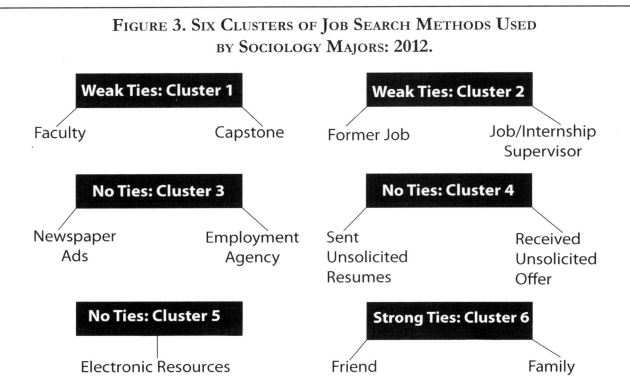

Figure 3. Six Clusters of Job Search Methods Used by Sociology Majors: 2012.

Source: American Sociological Association. *Social Capital, Organizational Capital, and the Job Market for New Sociology Graduates Survey*, 2012.

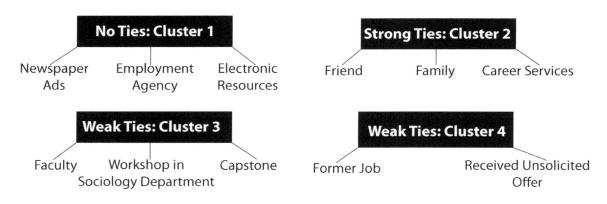

Figure 4. Four Clusters of Job Search Methods Used by Sociology Majors: 2012.

Source: American Sociological Association. *Social Capital, Organizational Capital, and the Job Market for New Sociology Graduates Survey*, 2012.

using classmates or friends and family. The use of career services was also in this cluster, although we would have expected it to cluster with weak ties. (In the previous six-cluster scale (Figure 3), career services did not fit into any of the clusters.) The third cluster in this analysis included what we have labeled as weak ties, that is, faculty members, workshops, and the capstone course. (The use of an internship supervisor as a job search strategy did not scale.) The final cluster included former employer and having received an unsolicited offer (see Figure 4). These two strategies did not appear to be of the same type even though they scaled. Perhaps the unsolicited offer came through a former employer. In general, the results of the cluster analysis did suggest that the categorization of job search strategies into types of ties were valid and did make sense, and we go ahead and use the 6-category factor analysis in the regression analysis that follows.

Box 1

Independent and Control Variables

1. Types of ties used in job search based on the two-variable cluster analysis, i.e. strong ties, weak ties, and impersonal ties.
2. Collapsed Carnegie codes for types of institutions of higher education.
3. Gender
4. Race or ethnicity
5. Father's education

Dependent Variable

Whether or not respondent obtained a career-type position.

REGRESSION ANALYSIS

We conducted a logit regression analysis to see what kinds of relationships and other factors, relative to one another, were predictive of gaining career-type positions (see Figure 5).[6] We used a logit model because the dependent variable had only two possible responses: yes or no. Rather than using individual items we used the results of the first cluster analysis as the independent variables, along with a series of control variables that included demographic characteristics: gender, race or ethnicity, type of undergraduate school, and parents' education. The most successful regression model contained the variables listed in Box 1.

The general lack of high correlations among the variables shows that multicolinearity was not a problem, therefore we were able to use all of the selected variables in the regression analysis.

The findings from the regression analysis generally support the descriptive analysis. Strong ties in the form of friends or relatives had a negative and statistically significant impact on the probability of a career position for sociology majors seven months after receiving their bachelor's degree, with those who used these ties only half as likely (.521) to have found career positions. As we have seen, those who used friends and family were more likely to find themselves in non-career positions such as clerical and administrative, sales, or service work. A major exception were those who had fathers with graduate education or graduate degrees, who likely had the social capital in the form of contacts and resources to have helped their children obtain professional-level positions. In this analysis, compared to the children of this group of better-educated fathers, the children of those who had less education were less likely to have found career-type jobs.

The use of all types of impersonal ties also had a negative and statistically significant impact on the probability of post-graduation sociology majors having found career-type jobs. These impersonal sources included the use of newspaper advertisements or employment agencies, resulting in respondents who used these strategies being 78% less likely to have found this kind of position. Those who sent unsolicited resumes and received unsolicited offers also had a lower probability of attaining career-type positions, although the use of this set of impersonal strategies was not statistically significant.

It was the use of weak ties that helped sociology

---

[6]We use the term predictive since the predictor (independent) variables come from the first wave of the study and the dependent variable comes from the second wave.

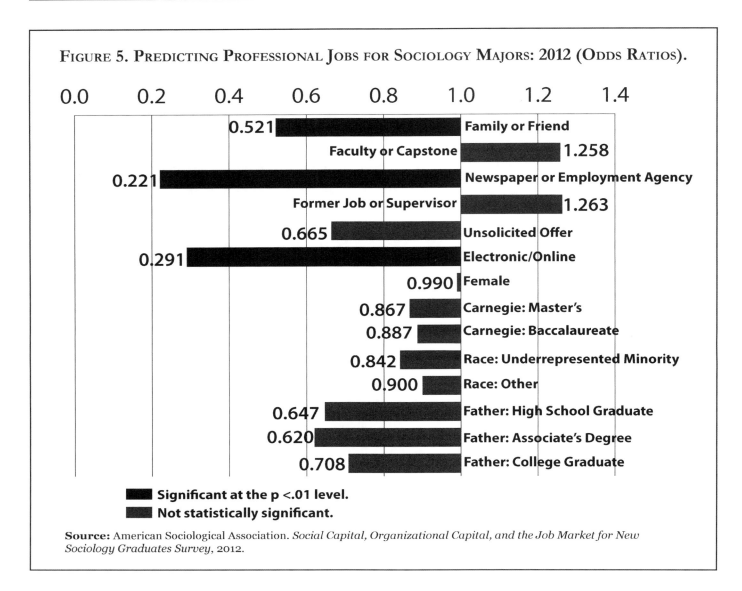

Figure 5. Predicting Professional Jobs for Sociology Majors: 2012 (Odds Ratios).

Source: American Sociological Association. *Social Capital, Organizational Capital, and the Job Market for New Sociology Graduates Survey*, 2012.

majors to obtain career-type positions such as social workers and career counselors, managers, IT and PR positions, researchers, and teachers. Two sets of weak relations had a positive impact on the probability of finding this type of position, although neither set was significant. The first set was help from faculty members and from capstone seminars (that increased the probability by 1.3 times), although we have seen that only about 11% of students used faculty members as part of a job search strategy and fewer than 2% of students participated in a capstone seminar. Calling upon internship supervisors or past employers was another set of weak ties that increased the probability of having found a career-type position (also by 1.3 times), although, here again, this finding was not statistically significant. In short, this analysis pointed in the direction of weak ties as a better method of finding career-type positions compared to strong ties and impersonal ties, since they take people out of their immediate circle of intimate relations and provide a set of social relations that can provide new information (Montgomery 1994).

SUMMARY AND CONCLUSIONS

Sociology majors from the class of 2012, who responded to our survey seven months after receiving their baccalaureate degrees, faced a tight job market and experienced close to a 10% unemployment rate. They were most likely to obtain social service/counseling jobs, although they obtained somewhat fewer career-level positions than did the class of 2005. This brief examined the use and effectiveness of three types of social ties and resources in their pursuit of career-level jobs, including strong, weak, and impersonal or non-ties.

We found that the largest group of respondents (56%) used impersonal ties in the form of online searches to look for jobs. This type of tie (or non-relation) along with the use of newspaper ads, employment agencies, and sending unsolicited resumes were not effective job search strategies for obtaining positions that the respondents designated as career-type jobs such as social service workers and counselors, managers, IT and PR professions, researchers, and teachers. Instead, impersonal ties were negatively related to these career-type positions, other factors being equal, and those who used these ties were more likely to have obtained clerical, sales, and service jobs.

Sociology majors may turn to these types of search strategies because they are readily available or seem on the face of it to be reasonable. However, this research suggests that graduates need to be mindful of the limitations of these strategies or may need to gain increased sophistication in the use of such strategies if they are to produce the reward of a career-type job as opposed to any job at all.[7]

Prior to the advent of the Internet and job search websites, available research suggested that the majority of job seekers used personal ties in their searches. The research was not always clear as to whether these sources of information were close ties or not. We found that the second-highest type of relation that sociology majors called upon were family members or classmates, colleagues, and friends (36% used them). These strong ties (with median scores of 5 or 4 on a 1-to-5 scale) were negatively related to working in career-type jobs, when other variables were held constant. Perhaps this negative effect was the result of the limitation of using close relatives and friends who did not take them beyond their immediate circle that could provide new information. We know that high percentages of sociology majors are first-generation college students, and majors whose family and friends work in clerical, sales, or service positions may not be able to provide graduates with information about jobs that are different from what they themselves have known. While not a significant difference, we did find data suggesting that majors with highly educated fathers are advantaged in the professional job market, presumably because of the resources and contacts that they are able to make available to their child.

Following researchers who found that weak ties were a more successful means of job searching than are strong ties, our results showed that weak ties in the form of advice from faculty members, capstone seminars, departmental career workshops, internship advisors, and former employers were positively related to the probability of obtaining a career-level job. However, these results were not statistically significant, perhaps because of the low percentage of sociology majors (only between 1% and 12%) who either had access to or used these resources. The relatively high percentages of respondents who believed they found a career level job who used these types of weak ties suggest that faculty members, sociology department activities, and outside activities such as internships may be undervalued as sources of job search information. Given the importance of sociology majors obtaining jobs that provide career ladders and that allow them to pay off their debts, sociology departments should consider promoting and increasing the kinds of activities and relations that result in weak job search ties.

Faculty meet with students on a regular basis. While they have neither the time nor inclination to become job counselors, faculty may want to use existing office hours and advising time to encourage students to begin to plan their job search strategies early in their undergraduate careers. And departments and faculty should explore ways of increasing knowledge about labor markets without having to expend inordinate amounts of time or resources doing so. For example, departments may want to enhance their ability to stay in touch with their alumni, for they may provide departments with job market information, speakers in capstone seminars, and sites for internships. Departments might also want to help increase the knowledge base of career service professionals so that they have a stronger understanding about the skills and knowledge base that sociology does provide students. Additional suggestions for faculty members can be found in *Launching Majors into Satisfying Careers: A Faculty Manual* (Spalter-Roth, Senter, and Van Vooren 2010).

---

[7] For example, Vitullo (2009) shows that using online searches with keywords such as "research" or "analysis" may yield more hits that the use of "sociologist."

APPENDIX I

Survey Design

First Wave

In this section we first describe the methodology used for Wave I of the study of the sociology cohort of 2012 that was conducted in the spring of their senior year. We go on to describe the methods used in Wave II of the survey, about seven months after respondents graduated. We review study research methods including sample selection, survey design, and data weighting.

Sample Selection

The 104 departments that participated in the 2005 *Bachelor's and Beyond* study were invited to take part in the 2012 study. Included in the 2005 group of departments was the stratified sample of 80 departments (20 from PhD granting institutions, 20 from MA institutions, and 40 from Baccalaureate institutions), as well as any additional volunteer departments. Along with matched departments, the 2012 invitation was extended to any department that wished to have its students included in the study. Departments were notified of the study via email, ASA's member newsletter *Footnotes*, and *Chairlink*, a listserv used to disseminate information to all depart¬ment chairs. The result was an additional 129 interested departments for a total of 233 departments that were expected to participate. Departments were asked to send a list of the names of their senior sociology majors graduating between April and August 2012, along with their email addresses. Departments that did not yet know who of their majors was graduating sent lists of all senior majors, and the response rate was later adjusted. Ultimately, 160 departments sent the ASA research department their lists after obtaining institutional review board (IRB) and/or any institutional approval necessary to disclose this information beyond the IRB approval granted to ASA by the Western Institutional Review Board (WIRB).

Questionnaire and Responses

ASA's Department of Research on the Discipline and Profession created the student survey, along with the help of the study's Advisory Committee. The 2012 survey replicated many questions from the 2005 questionnaire, with additional questions about the social networks students used or planned to use in pursuing jobs or admission into graduate school. Questions focused on students' experiences as sociology majors, including why they majored, skills and concepts they learned, activities they participated in, their job and graduate school aspirations, and the types of relations used in finding appropriate jobs and graduate schools. Respondents were expected to use an online version of the survey, created by Indiana University's Center for Survey Research. The survey was pre-tested in November 2011 by the Advisory Committee members' students. The final version of the survey was launched with an invitational email to students in March 2012, which was followed up with four reminder emails before the survey closed in early May. By the time it closed, 2,695 students had participated in the survey, for an average departmental response rate of about 40 percent (36.8%).

Weighting

The 2012 Wave I data were weighted so that they are more reliable. We compared demographic and institutional characteristics of respondents with those of recent baccalaureates in sociology, based on the National Center for Educational Statistics Integrated Postsecondary Education Data system (IPEDS). These characteristics included gender, race and ethnicity, and type of institution of higher education. We had weighted the 2005 data by institution type. In 2012, there were only small differences by race or ethnicity and by institution type compared to the IPEDS data. The largest underrepresentation was seen among Black/African American respondents. To adjust for this in the 2012 data only, we coded anyone who selected black into Black/African American, even if they also selected another racial category. In addition, there was a disparity between the percentage of male graduates and the percentage of male respondents. Therefore, we weighted the responses by gender. However, when we present comparison findings from 2005 with 2012 we use unweighted data since the comparative data are placed in a single dataset in which the 2005 and the 2012 weights cannot both be applied. The differences between the unweighted and the weighted data are relatively small.

Second Wave

Sample Universe, Survey, and Response Rates

In the fall of 2012, the Research Department with the aid of the project's advisory committee[8] developed the questionnaire for the second wave of the 2012 survey. The 2012 survey replicated many questions from the 2005 questionnaire, was approved by the Western Institutional Review Board (WIRB), and was developed as an online survey by the Indiana University Center for Survey Research. Once the survey was complete and pre-tested, a letter of invitation was sent. All of the 2,695 respondents who participated in the first wave of the study were sent the 15 minute survey that asked questions about their employment and/or graduate school status, after their 2012 graduation. They were asked about the characteristics of their jobs; the ties (that according to our measures could be labeled as strong, weak, or non-personal), sources that were most helpful in their job or graduate school search; and which skills and concepts they used on the job or in graduate school. The survey concluded with a few additional questions about current living and family situations. The survey was sent out at the beginning of December 2012, with reminders sent to what students listed as their primary and secondary email addresses, between mid-December 2012 and the end of January 2013. The survey was closed down in February of 2013. The total response rate was just over 41% (with 1,108 completed surveys). An additional 62 respondents returned partially completed surveys, with 127 "not deliverables" and 15 refusals for a refined response rate of just over 44%—a higher response rate than for the first survey.

Weighting

When we examined the response rates of each of the sub-groups (type of institution of higher education, race and ethnicity, and gender) in the sample of respondents, we did not find significant differences between the response rates to this wave of the survey and their proportion in the sample. Therefore, we did not weight the second wave.

REFERENCES

Baum, Sandy and Michael McPherson. 2010. "Gainful Employment." *The Chronicle of Higher Education.* Washington, DC. Retrieved September 10, 2010 (http://chronicle.com/blogPost/GainfulEmployment/26770/).

Bourdieu, P. 1986. "The Forms of Capital." In Richardson, J. G. (ed.), *Handbook of Theory and Research for the Sociology of Education*: 241-258. New York/Westport/London.

Brint, Steven. 2002. "The Rise of the 'Practical Arts'." In *The Future of the City of Intellect: The Changing American University*: 231-259. Stanford, CA: Stanford University Press.

Brint, Steven. 2005. "Guide for the Perplexed: On Michael Burawoy's 'Public Sociology'." *The American Soci-*

---

[8] The Advisory Committee consists of the project co-PI Mary Senter of Central Michigan University, Margaret (Peggy) Nelson of Middlebury College, John Kennedy of Indiana University, Pamela Stone and Michael Wood of Hunter College, City College of New York, and Jeffrey Chin of LeMoyne College.

ologist Fall/Winter:46-65.

Brint, Steven. 2010. "Sociology and the Market-Model University." American Sociological Association's Chairs Conference, Atlanta, GA.

Carnevale, A. P., Cheah, B., and Strohl, J. 2013. *Hard Times: College Majors, Unemployment, and Earnings.* Washington D.C.: Georgetown Center on Education and the Workforce, Georgetown Public Policy Institute.

Fernandez, R.M., Castilla, E.J., and Moore, P. 2000. "Social Capital at Work: Networks and Employment at a Phone Center." *American Journal of Sociology* 105(5):1288-1356.

Gardner, Lee. 2013. "College Leaders Strive for Performance Measures that Fit their Institutions." *Chronicle of Higher Education.* July 10. Retrieved July 11, 2013 (http://chronicle.com/article/College-Leaders-Strive-for/140159/).

Grannis, Rick. 2010. "Six Degrees of 'Who Cares?'" *American Journal of Sociology* 115:991-1017.

Granovetter, Mark. 1995. *Getting a Job: A Study of Contacts and Careers 2nd ed.* Cambridge, MA: Harvard University Press.

Granovetter, Mark. 1983. "The Strength of the Weak Tie: Revisited." *Sociological Theory* 1: 201-33.

Granovetter, Mark. 1973. "The Strength of Weak Ties." *American Journal of Sociology* 78:1360–80.

Kirschenman, Joleen and Kathryn Neckerman. 1991. "'We'd Love to Hire Them, But . . .': The Meaning of Race for Employers," Pp 203–232 in *The Urban Underclass* edited by Christopher Jencks and Paul Peterson. Washington DC: Brooking Institute.

Krackhardt, D. (1992). "The Strength of Strong Ties: The Importance of Philos in Organizations." In N. Nohria & R. Eccles (eds.), *Networks and Organizations: Structure, Form, and Action*: 216-239. Boston, MA: Harvard Business School Press.

Kuh, George, and Hu, S. (2001). "The Effects of Student-faculty Interaction in the 1990s." *Review of Higher Education* 24: 309–332.

Lin, Nan. (2001). *Social Capital: A Theory of Social Structure and Action.* Cambridge: Cambridge University Press.

Marsden, Peter V. and Elizabeth H. Gorman. 2001. "Social Networks, Job Changes, and Recruitment." Pp. 467–502 in *Sourcebook of Labor Markets: Evolving Structures and Processes,* edited by Ivar Berg and Arne L. Kalleberg. New York: Kluwer Academic/Plenum.

Martin, Nathan D. (2009). "Social Capital, Academic Achievement, and Postgraduation Plans at an Elite, Private University." *Sociological Perspectives,* 52(2): 185-210.

McDonald, Steve and Glen H. Elder Jr. 2006. "When Does Social Capital Matter? Non-Searching For Jobs Across the Life Course." *Social Forces* 85(1):521-48.

McPherson, M., Smith-Lovin, L., and Cook, J.M. (2001). "Birds of a Feather: Homophily in Social Networks." *Annual Review of Sociology* 27: 415-444.

Montgomery, J.D. (1994). "Weak Ties, Employment, and Inequality: An Equilibrium Analysis," *American Journal of Sociology* 99 (Mar.): 1212-36.

Mouw, Ted. 2003. "Social Capital and Finding a Job: Do Contacts Matter?" *American Sociological Review* 68:868-98.

Neckerman, Kathryn M. and Roberto F. Fernandez. 2003. "Keeping a Job: Network Hiring and Turnover in a Retail Bank." *Research in the Sociology of Organizations* 20:299-318.

O'Reagan, Katherine and John Quigley. 1993. "Family Networks and Youth Access to Jobs." *Journal of Urban Economics* 34:230–48.

Pascarella, Ernest T. and Patrick T Terenzini. (2005). *How College Affects Students: A Third Decade of Research*. The Jossey-Bass higher and adult education series. San Francisco: Jossey-Bass.

Reskin, Barbara. 1998. *The Realities of Affirmative Action in Employment*. Washington, DC: American Sociological Association.

Rosenbaum, J.E., DeLuca, S., Miller, S.R., and Roy, K. 1999. "Pathways into Work: Short- and Long-Term Effects of Personal and Institutional Ties." *Sociology of Education* 72:179-96.

Senter, Mary S., Nicole Van Vooren, Michael Kisielewski, and Roberta Spalter-Roth. 2012. *What Leads to Student Satisfaction with Sociology Programs?* Washington, D.C.: Department of Research and Development on the Discipline and Profession, American Sociological Association.

Spalter-Roth, Roberta, Mary Senter, Michael Kisielewski, and Nicole Van Vooren. 2013. *Recruiting Sociology Majors: What Are the Effects of the Great Recession? Concepts, Change, and Careers*. Washington, DC: American Sociological Association.

Spalter-Roth, Roberta. Mary Senter, and Nicole Van Vooren. 2010. *Launching Majors into Successful Careers: A Faculty Manual*. Washington, D.C., The American Sociological Association. Retrieved July 11, 2013 www.asanet.org/documents/research/pdfs/ASA_Launching_Majors_Faculty_Manual_2010.pdf.

Spalter-Roth, Roberta and Nicole Van Vooren 2010. *Mixed Success: Four Years of Experiences of the Class of 2005*. Washington, D.C., The American Sociological Association. Retrieved July 11, 2013 www.asanet.org/research/BBMixedSuccessBrief.pdf.

Spalter-Roth, Roberta and Nicole Van Vooren. 2008. *Pathways to Job Satisfaction: What Happened to the Class of 2005*. Washington, D.C.: Department of Research and Development on the Discipline and Profession, American Sociological Association.

Vitullo, Margaret. 2009. "Searching for a Job with an Undergraduate Degree in Sociology."American Sociological Association: *Footnotes*. 37 (7) (September/October. Retrieved July 11, 2013 www.asanet.org/footnotes/septoct09/index.html.

Wegener, Bernd. 1991. "Job Mobility and Social Ties: Social Resources, Prior Job, and Status Attainment." *American Sociological Review* 56:60-71.

The following are selected research briefs and reports produced by the ASA's Department of Research on the Discipline and Profession for dissemination in a variety of venues and concerning topics of interest to the discipline and profession. These and all research briefs are located at www.asanet.org/research/briefs_and_articles.cfm. You will need Adobe Reader to view our PDFs.

Title	Format	Year
The Victory of Assessment? What's Happening in Your Department?: The AY 2011-2012 Department Survey	PDF	2013
Changes in Technology, Courses, and Resources: What's Happening in Your Department?: The AY 2011-2012 Department Survey	PDF	2013
Postdocs: Another Stage in the Sociology Pipeline?	PDF	2013
Sociology Majors: Before Graduation in 2012	PDF	2013
Recruiting Sociology Majors: What Are the Effects of the Great Recession?: Concepts, Change, and Careers	PDF	2012
What Leads to Student Satisfaction with Sociology Programs?	PDF	2012
On the Upswing: Findings from the ASA 2011-2012 Job Bank Survey	PDF	2012
What Do We Know About the Dissemination of Information on Pedagogy?: 2008, 2010, and 2011	PDF	2012
Mothers in Pursuit of Ideal Academic Careers	PDF	2012
Research about Minorities in Sociology: Surveys, Datasets, and Measurement	PPT	2012
The Effects of New Technology on the Growth of a Teaching and Learning Network	PDF	2011
The Future of Sociology: Minorities, Programs, and Jobs	PPT	2011
The Impact of Cross Race Mentoring for "Ideal" and "Alternative" PhD Careers in Sociology	PDF	2011
Sociology Master's Graduates Join the Workforce	PDF	2011
Are Masters Programs Closing? What Makes for Success in Staying Open?	PDF	2011
Falling Behind: Sociology and Other Social Science Faculty Salaries, AY 2010-2011	PDF	2011
A Decade of Change: ASA Membership From 2000 - 2010	PDF	2011
Findings from ASA Surveys of Bachelor's, Master's and PhD Recipients	PDF	2011
Homosociality or Crossing Race/Ethnicity/Gender Boundaries? Pipeline Interventions and the Production of Scholarly Careers	PDF	2011
Networks and the Diffusion of Cutting-Edge Teaching and Learning Knowledge in Sociology	PDF	2010
The Gap in Faculty Pay Between Private and Public Institutions: Smaller in Sociology than in Other Social Sciences	PDF	2010
From Programs to Careers: Continuing to Pay Attention to the Master's Degree in Sociology	PDF	2010
Teaching Alone? Sociology Faculty and the Availability of Social Network	PDF	2010

Follow the Department of Research on Facebook at http://www.facebook.com/ASAResearchDepartment and on Twitter at https://twitter.com/ASAResearch

American Sociological Association
Department of Research on the Discipline and Profession
www.asanet.org
research@asanet.org

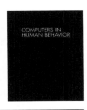

Computers in Human Behavior

journal homepage: www.elsevier.com/locate/comphumbeh

Facebook fired: Legal perspectives and young adults' opinions on the use of social media in hiring and firing decisions

Michelle Drouin [a,*], Kimberly W. O'Connor [b], Gordon B. Schmidt [b], Daniel A. Miller [a]

[a] *Department of Psychology, Indiana University–Purdue University, Fort Wayne, United States*
[b] *Department of Organizational Leadership and Supervision, Indiana University–Purdue University, Fort Wayne, United States*

ARTICLE INFO

Article history:
Available online 24 January 2015

Keywords:
Social networking
Young adults
Job terminations
Psychological characteristics
Social media

ABSTRACT

In this exploratory study, we examine young adult undergraduates' (n = 448) opinions regarding the use of social media for employment decisions, a practice that has been highlighted in the popular press and recent legal cases. Most of the young adults in our sample were not in support of this practice (only one third were), and most expressed a liberal view of what should be permissible for posting on social media without the threat of job termination (e.g., less than half believed that posting illegal sexual behavior online should result in termination). Additionally, those who were most opposed to using social media in employment decisions were older, had less self-control, were more endorsing of the hookup culture, and were more open to experience. We discuss these findings with regard to current social media/work life issues, suggesting that: (1) these opinions may affect companies and legal entities who are developing social media policies, but also (2) that young adults need to be aware that regardless of their opinions on the practice, their social media use could have long-term effects on their careers.

© 2015 Elsevier Ltd. All rights reserved.

1. Introduction

In a recent Pew survey, 73% of American adults reported using the internet to engage in social networking (Duggan & Smith, 2014). Among young adults (18–29), the percentage of users is even higher: Recent surveys show that 83% of young adults use social media sites (Duggan & Brenner, 2013). And trends show no signs of a social media slow-down. In 2013, more American adults were using the most popular social networking sites (i.e., Facebook, Twitter, Instagram, Pinterest, and LinkedIn) than had in 2012, 42% of internet users indicated they used at least two of the most popular sites, and many visited the sites daily (Duggan & Smith, 2014).

Research indicates that the motivation behind individuals' use of social media is often to develop or maintain social or romantic relationships, to feel connected to others, to gain information, or to gain social capital (Bonds-Raacke & Raacke, 2010; Gangadharbatla, 2008; Nadkarni & Hofmann, 2012; Sheldon, Abad, & Hinsch, 2011; Smith, 2011). Because of the range of motivations, diversity of social relationships, and the vast quantity of subscribers, personal social media usage has inevitably become intertwined with the workplace. One recent study indicated that 60% of employees report having one or more co-worker "friends" on Facebook, and 25% of employees report that they are Facebook friends with their supervisor (Weidner, Wynne, & O'Brien, 2012). Significant issues therefore arise when content that is not professional is seen by supervisors, co-workers, or other types of professional contacts. Those issues can result in serious consequences for social media users, as well as the organizations that employ them, as evidenced by the growing number of employment disputes related to social media that have resulted in litigation.

In this exploratory study, our goals were to: (1) measure young adults' perceptions of fairness of employers' use of social media for making employment decisions, and (2) examine how personality, individual differences, and personal social media use impacts such fairness perceptions. To contextualize these findings, we include a legal perspective on the social media and work cases that have emerged in the courts, and then focus our discussion on the implications of these results for employees and organizations, emphasizing the effect these perceptions may have on future law and policy.

1.1. Social media's increased role in employment decisions: a legal perspective

Social media has had an increasingly significant impact on human resource practices and has been the basis for many recent employment decisions that have resulted in litigation. A 2013 study by Jobvite found that 93% of recruiters said they were likely

* Corresponding author at: Indiana University–Purdue University, Fort Wayne 2101 E., Coliseum Blvd., Fort Wayne, IN 46805, United States.
E-mail address: drouinm@ipfw.edu (M. Drouin).

http://dx.doi.org/10.1016/j.chb.2015.01.011
0747-5632/© 2015 Elsevier Ltd. All rights reserved.

to look at the social media profiles of applicants, and 43% have reconsidered a candidate (both in the negative and positive direction) based on the candidates' social media profile (Jobvite, 2013). With regard to existing employees, 17% of organizations report they have had issues with employees' use of social media, and 8% say they have actually dismissed a worker for a social media behavior (Ostrow, 2009).

Public opposition to this type of scrutiny often occurs when an organization fires an employee for social media conduct that appears to be unrelated to the workplace. For example, high school English teacher, Ashley Payne, was asked to resign after she posted a picture from her European summer vacation on Facebook. The picture depicted Payne smiling and holding two drinks while in an Irish pub (Payne v. Barrow County School District & (Super. Ct. Ga., 2009). Meanwhile, middle school teacher, Anna Land, was fired after a picture of her was posted (and subsequently taken down) by an unknown third party. The picture showed Land in a simulated sexual act with a male mannequin while at a bachelorette party (Land v. L'Anse Creuse, 2010). These are just two of a number of incidents where employees have been fired for posts that have been placed on social media sites. The popular press has coined the term, "Facebook Fired," to refer to the growing number and type of incidents that have arisen across all professions (Hidy & McDonald, 2013). However, K-12 public school teachers, such as Ashley Payne and Anna Land, have been particularly hard hit as a profession, since the general perception is that teachers should be held to a higher moral standard than other types of professionals (Fulmer, 2010; McNee, 2013; Miller, 2011).

There is very little legal protection afforded to teachers' online communication under the First Amendment in these cases. As public sector employees, the long-standing legal standard is that their speech is protected from adverse employment action only if it involves a matter of "public concern" (matters of political, social, or other concern to the community) (Pickering v. Bd. of Ed., 1968). Meanwhile, private sector employees have almost no First Amendment protection from employer discipline for their online communications, though some protection to discuss the terms and conditions of their employment comes from the National Labor Relations Act (Fulmer, 2010; Raphan & Kirby, 2014).

Negative reaction from the public has resulted in some legal protection for employees by way of state law. For example, a common practice had emerged where employers were asking employees' and/or recruits for their social media usernames and passwords (McFarland, 2012). Legislatures in nearly twenty states have recently enacted laws to prohibit employers from engaging in this behavior, and many more states have legislation pending on this issue (Stinson, 2014). Also, some states are now requiring school corporations to implement social media policies in an effort to define and curtail "inappropriate" teacher online activity (DiMarzo, 2012). Though some other types of employers have developed crucial social media policies to deal with issues involving employee's social networking practices, studies show that 57% have not (Mulvey, 2013). Many personal social media usage issues therefore remain undefined for employees as they relates to their workplaces (Protivity, 2013). The laws and policies regarding social media-based terminations of employment are still evolving and are being shaped by societal notions of justice and fairness.

1.2. Existing research on social media, psychological characteristics, and employment decisions

In the empirical literature on the topic, there are currently no known studies that have measured societal opinions of the use of social media for employment decisions. However, a few researchers have attempted to connect social media use, personality characteristics, and employment-related outcomes. In one of the first studies in this area, Karl, Peluchette, and Schlaegel (2010), surveyed American and German undergraduates to determine the frequency with which young adults posted problematic material (e.g., drug and alcohol use or sexual behavior) online and how this related to individual personality characteristics. Karl et al. (2010) found that a fair number of young adults, more Americans than Germans, indicated that they had posted different types of problematic material online. Undergraduates who were more conscientious, agreeable, and emotionally stable were less likely to post problematic material online, whereas those who had greater internet compulsivity were more likely to post such material online (Karl et al., 2010). In the discussion of their results, Karl et al. (2010) connected their findings to employment decisions, suggesting that these internet profiles may be used by employers to make hiring decisions during the recruitment and selection processes.

Stoughton, Thompson, and Meade (2013) took this work a step further by examining the relationships between personality characteristics and potentially damaging online behaviors in a pool of actual job applicants. In their study, college students who were more extraverted were more likely to reference alcohol and drugs in posts. Additionally, those who were less agreeable were more likely to badmouth others (e.g., coworkers, classmates, professors, or superiors) in posts; whereas those who were more conscientious were less likely to badmouth others (Stoughton et al., 2013). Stoughton et al.'s (2013) study provided further evidence that there is a link between a person's psychological characteristics and their potentially damaging online behavior. Again, these authors connected their findings to employment decisions, suggesting that some employers may already use social media sites to screen potential applicants, and that Facebook profiles may have the potential to provide valuable information to employers about the psychological characteristics of job applicants.

Notably, although both Karl et al. (2010) and Stoughton et al. (2013) suggested that social media profiles might be useful to employers to help make decisions about job applicants, neither examined applicants' opinions about this process. Buy-in from job applicants and employees on these issues may lessen the likelihood of litigation; therefore, it is important from both a practical and legal perspective to understand what these opinions are and which types of individuals hold these opinions.

1.3. Opinions of use of social media for employment decisions: the current study

As social media is being used in employment decisions (e.g., Jobvite, 2013), but social media policies are still mostly non-existent or in flux (Mulvey, 2013), public opinions about these practices may help to shape future policy. Opinions about the appropriateness of using social media profiles in employment decisions are likely shaped by both individual characteristics, like the Big-5 characteristics examined in Karl et al.'s (2010) and Stoughton et al.'s (2013) studies, as well as societal factors, such as the proliferation of social media use in the culture.

With regard to Big-5 traits, we expected that psychological openness would affect opinions on the use of social media for employment decisions. Psychological openness is one of the five key personality dimensions according to the five factor model (e.g., Costa & McCrae, 1992; Digman, 1990). McCrae (1996) defines openness as a broad but complex personality dimension which encompasses both intrapsychic and interpersonal dimensions. Those who are high in openness to experience are creative and flexible thinkers, who are open to new experiences and are intellectually curious (McCrae, 1996). With regard to the present inquiry, we hypothesized [H1] that those high in openness would be more likely to *oppose* the use of social media for hiring and firing decisions, as they would be more likely to accept a wider variety of behaviors (both online and offline) as acceptable within the spectrum of normal human behavior.

Aside from psychological openness, we also expected a few other characteristics to be related to the opposition of the use of social media for employment decisions, including the use of and addiction to social media, low self-control, and endorsement of the hookup culture. With regard to social media use, we expected those who use social media to a greater extent to be more likely to have posted things online that could be potentially damaging to their career or job search. This might be even more apparent among those who are addicted to Facebook, who, according to Andreassen, Torsheim, Brunborg, and Pallesen (2012) may have "used Facebook so much that it has had a negative impact on [their] job/studies" (p. 516). Indeed, Karl et al. (2010) found that those who had higher internet compulsivity were more likely to have posted potentially problematic material online. Thus, we expected [H2] that because they had more opportunities to post potentially damaging information online, those with high social media use and Facebook addiction would be more opposed to the use of social media for hiring and firing decisions.

We also expected that those with low self-control would oppose the use of social media for employment decisions. Researchers have shown that those with low levels of self-control are more likely to engage in inappropriate or socially non-normative behaviors (DeBono, Shmueli, & Muraven, 2011), including inappropriate or impulsive sexual acts (Gailliot & Baumeister, 2007). Moreover, according to Vohs, Ciarocco, and Baumeister (2005), when self-control is lower, individuals are less effective at impression management. Therefore, we expected those with low self-control would have both engaged in more impulsive behaviors and been less effective at impression management (e.g., through their online posts). Consequently, akin to those who use social networking frequently, we expected [H3] those with low self-control to have more potentially damaging information online and therefore oppose the use of social media for employment decisions.

Finally, we explored a variable that reflects acceptance of one of the normative values of modern young adult culture—endorsement of the hookup culture (Aubrey & Smith, 2013). According to Aubrey and Smith (2013), those who endorse the hookup culture do not believe that young adulthood is a time for commitment; instead, they embrace the fun, sexually-free nature of the life stage, and they engage in sexual hookups to attain status and/or maintain control. Therefore, like those low in self-control, they may have engaged in more potentially damaging behaviors that could have been posted online. Alternatively, like those high in openness to experience, they may be more flexible thinkers, who would be more open to different lifestyles, with the understanding that these lifestyles may have been documented (inappropriately) online. It might also be that those who endorse the hookup culture are part of a larger movement that embraces the norms (e.g., social networking, hashtagging, tweets) of the young adult generation and rejects the policies and laws imposed upon them by their elders.

Regardless of the reason, we expected [H4] a positive relationship between endorsement of the hookup culture and opposition to the use of social media for employment decisions.

2. Methods

2.1. Participants

Participants were 442 young-adult students (144 men, 298 women) from a mid-sized Midwestern university. From a larger sample ($N = 477$), only the young adult participants (aged 18–27) were retained. The participants' average age was 19.45 years ($SD = 1.67$) and most were freshmen (65% freshmen, 25% sophomores, 6% juniors, and 3% seniors). Participants came from more than 40 major fields of study. With regard to ethnicity, most of the sample was Caucasian (84% White, Non-Hispanic; 5% Hispanic; 3% African American; 3% Asian; and 5% Biracial or other ethnicity).

2.2. Procedure

Participants were recruited from introductory psychology classes in Fall, 2013 and received a research credit for participation. All participants completed online consent forms and were then given access to an online anonymous survey on social media use among young adults. The survey included demographic questions as well as questions about their opinions on the use of social media for hiring and firing decisions, their use of and addiction to social media, and various psychological measures. As this was part of a larger study, only the relevant measures are reported below.

2.3. Measures

2.3.1. Opinions of use of social media for employment decisions

To measure participants' opinions about the use of social media for employment decisions, we asked them to respond on a 5-point agreement scale (1 = *strongly disagree*, 5 = *strongly agree*) to the following statement: "A person's Facebook or Twitter account should not be used to make hiring or firing decisions." As this study was exploratory in nature, we also asked them a number of other questions related to this practice (see Table 1 for questions), including their agreement with the legal outcomes of recent social media cases (questions 2 & 3) and questions related specifically to illegal lewd behavior, which was defined as "sexual gratification with another with knowledge that they are in the presence of others OR publicly and indecently exposing genitals or pubic area" (questions 3 & 4). We also asked them whether they feared whether something they posted might hurt them in their job search (question 5). For each of these questions (displayed in Table 1), participants responded on a 5-point agreement scale (1 = *strongly disagree*, 5 = *strongly agree*). Additionally, to measure whether they or their friends had ever felt the threat of job loss from social media, we asked two separate questions on the extent to which:

Table 1
Participants' opinions and feelings about social media and hiring and firing decisions.

	Disagree (%)	Neutral (%)	Agree (%)
1. A person's Facebook or Twitter account should not be used to make hiring or firing decisions	30	28	42
2. It is acceptable for a teacher (k-12) to post a picture on her Facebook site of her holding a beer during a vacation to Ireland	24	24	52
3. If a teacher (k-12) engages in lewd behavior [legal definition provided] at a bachelor or bachelorette party and someone posts and tags pictures of that teacher online, the teacher should lose his or her job	39	28	33
4. People should be able to post pictures of private events (e.g., parties) without threat of losing their job, even if those pictures contain lewd behavior [legal definition provided]	44	27	29
5. I fear that some pictures/videos posted of me will hurt me in my job search	78	12	10

Note. Disagree = strongly disagree or disagree. Agree = strongly agree or agree.

(1) the person had ever "Lost job/thought I might lose job because of something posted online," (2) a friend had ever "Lost job/thought they might lose job because of something posted online." For these questions, participants responded on a 5-point frequency scale (1 = *never*, 5 = *very frequently*).

2.3.2. Openness to experience

Openness to experience was measured with the openness to experience subscale of the 44-item Big Five Inventory (John & Srivastava, 1999). Participants indicated their level of agreement of whether they saw themselves as someone who, for example, "Is curious about many different things" on a 5-point Likert scale (1 = *disagree strongly*, 5 = *agree strongly*). Cronbach's alpha for this subscale was .74.

2.3.3. Frequency of social media use

Participants were asked how frequently they posted messages or sent media via five of the most popular social communication methods: Text message, Facebook, Twitter, Instagram, and Snapchat. Participants responded on a 6-point Likert scale (1 = *never*, 6 = *very frequently*). These five measures were combined into one composite measure reflecting the frequency of social media use (Cronbach's alpha = .71).

2.3.4. Facebook addiction

Facebook addiction was measured with the 6-item Bergen Facebook Addiction Scale (Andreassen et al., 2012), which asked participants to respond on a 5-point Likert scale (1 = *very rarely*, 5 = *very often*) on the frequency with which they had engaged in addictive Facebook behaviors (e.g., "Felt an urge to use Facebook more and more.") during the last year. Cronbach's alpha for this measure was .88.

2.3.5. Self-control

To measure self-control we used the 13-item short form of the Self-Control Scale (SCS; Tangney, Baumeister, & Boone, 2004). Participants were asked to respond about how well statements (e.g., "I am good at resisting temptation.") reflected how they typically are on a 5-point Likert scale (1 = *not at all*, 5 = *very much*). Cronbach's alpha for this measure was .84.

2.3.6. Endorsement of the hookup culture

Attitudes towards the modern-day hookup culture were measured with the 20-item Endorsement of the Hookup Culture Index (EHCI; Aubrey & Smith, 2013). Participants indicated their agreement with statements such as "Hooking up is not a big deal" on a 5-point Likert scale (1 = *strongly disagree*, 5 = *strongly agree*). Cronbach's alpha for this subscale was .97.

3. Results

As this study was exploratory in nature, our first step was to examine the frequencies with which people engaged in behaviors and supported beliefs that social media should be used for employment decisions. Next, we conducted a correlational analysis to examine whether our social communication and psychological measures were related to the belief that social media should be used for employment decisions. Finally, we conducted a hierarchical regression analysis, controlling for age and gender, to determine whether our psychological characteristics of interest contributed unique variance to this belief.

With regard to actual and perceived threat of job loss related to social media, only 6% of participants lost a job or thought they might lose their job based on their social media posts, but 25% indicated that they had a friend who had lost their job or thought they might lose their job based on their social media posts. In terms of their opinions regarding social media and employment decisions, 42% believed that social media should *not* be used for hiring and firing decisions (see Table 1). With regard to their attitudes toward actual court cases, the majority of participants (53%) agreed that it was acceptable for a teacher to post a picture of herself holding a beer on a social networking site. In terms of illegal, lewd behavior, 39% of participants (and 27% were neutral) *disagreed* that a teacher who had pictures posted of her engaging in lewd behavior during a bachelorette party should be fired. Meanwhile, in terms of general lewd behavior, 29% agreed (and 27% were neutral) that lewd behavior posted from private events should *not* cost a person their job. Thus, overall, approximately two-thirds of participants were either neutral or positive (in terms of keeping a job) towards people who were featured in lewd acts online. Finally, only a small portion of the sample (10%) feared that something they had posted might hurt them in their job search.

Our next step was to examine whether there were any significant relationships between our outcome measure of interest (Item #1 from the table—young adults' opinions of whether social media should be used for hiring and firing decisions), social media usage and addiction, and our selected personality measures. We also included age and gender in the analyses as potential covariates. Our correlational analyses showed that there were significant but weak relationships between agreement that social media should *not* be used for hiring or firing decisions and self-control, openness to experience, and endorsement of the hookup culture (Table 2). In contrast, agreement that social media should not be used for hiring or firing was not significantly related to overall frequency of social media use or Facebook addiction. Therefore, our H1, H2, and H4 were supported, but H3 was not supported. Meanwhile, with regard to our potential covariates, age and gender were both positively and significantly related to the opposition of the use of

Table 2
Relationships between agreement that social media should not be used for hiring/firing, social media use, and personality characteristics.

	1	2	3	4	5	6	7	8
1. No SM for hiring/firing	–							
2. Age	.13**	–						
3. Gender	.12**	.18**	–					
4. BFI-Openness	.16**	.25**	.10*	–				
5. Frequency SM use	.03	−.18**	−.25**	−.09*	–			
6. FB addiction	.02	.02	−.16**	−.04	.22**	–		
7. Self-control	−.16**	−.03	−.15**	.08	−.15**	−.18**	–	
8. EHCI	.21**	.20**	.47**	.01	.06	.02	−.42**	–
M	3.20	19.45	0.33	3.51	1.80	3.36	3.84	2.20
SD	1.25	1.67	0.47	0.59	0.81	0.65	1.18	1.00

Note. SM = social media. BFI = Big Five Inventory. EHCI = Endorsement of the Hookup Culture Index. For gender, men = 1, women = 0.
* $p < .05$.
** $p < .001$.

Table 3
Hierarchical regression of agreement that social media should not be used for hiring or firing on self-control, openness to experience, and endorsement of hookup culture.

Variable	B	SE(B)	β	ΔR²
Step 1				.03**
Age	.09	.04	.12*	
Gender	−.27	.12	.09	
Step 2				.05***
BFI-Openness	.33	.10	.15**	
Self-control	−.20	.10	−.10*	
EHCI	.19	.07	.15*	

Note. Total $F(5, 425) = 7.70$, $R^2 = .08$. For gender, men = 1, women = 0.
* $p < .05$.
** $p < .001$.
*** $p < .001$.

social media for hiring and firing; therefore they were included as control variables in the first step of our regression analyses that followed.

Next, we performed a hierarchical regression analysis, controlling for age and gender in the first step, to determine whether self-control, openness to experience, and endorsement of the hookup culture were unique predictors of agreement that social media should not be used for hiring or firing (see Table 3). In the first step, age was a significant predictor: Those who were older were more likely to agree that social media should not be used in hiring or firing decisions. In the next step, after controlling for age and gender, openness to experience, self-control and endorsement of the hookup culture were all significant predictors of agreement that social media should not be for these purposes. Those who were more open to experience, had less self-control, and endorsed the hook-up culture more were more likely to agree that social media should not be used for hiring or firing decisions. Overall, our predictors accounted for 8% of the variance in this belief.

4. Discussion

Social media use has skyrocketed over the last decade, and companies and legislative bodies are scrambling to develop policies to address the myriad of issues that have arisen as a result of the inevitable mixing of work and private life through this medium. The aim of our exploratory study was to examine whether young adults agreed with the use of social media for employment decisions (e.g., hiring and firing), whether they agreed with the outcomes of recent legal cases on the topic, and what views they held on the posting of illegal material with regard to job termination. Additionally, we examined the psychological characteristics of those who opposed the use of social media for hiring and firing decisions.

Almost half of the participants (44%) stated that social media should *not* be used for used in hiring and firing decisions, and a further 28% were neutral. Therefore, less than one third of our young adult sample expressed agreement with this practice. Meanwhile, with regard to their opinions about behaviors featured in recent legal cases, only one fourth of our participants deemed a k-12 teacher holding a beer in a social media picture to be unacceptable, and only one third agreed that a k-12 teacher featured on social media engaging in lewd behavior should lose her job. When questioned more generally about the practice of posting illegal sexual behavior online, almost one third (29%) of the young adults in our sample felt that people should be able to post *illegal* material without the threat of losing their jobs. Moreover, only 10% feared that something that they had posted could hurt them in their job search. Overall, these results suggest a very liberal view of the types of material that people should be able to post online without the threat of job termination. More importantly, they show that this generation (young adults) generally do not support employers' use of social media for making employment decisions about their employees.

Those who most opposed the practice of using social media for employment decisions had less self-control, were more likely to endorse the hookup culture, and were more psychologically open. These results were expected. We hypothesized that those with these psychological characteristics (i.e., less self-control, more openness to experience, and more acceptance of the hookup culture) may have more opportunities to engage in inappropriate behavior and may have more to hide on their social media profiles. Thus, their opposition of the use of social media for job decisions could be a form of self-preservation. Alternatively, we suggested that those with these characteristics may have a more liberal view and flexibility with regard to what is considered normal within the context of human behavior. The fact that all of these psychological characteristics emerged as unique predictors of the opinion of use social media use for hiring and firing decisions suggests that both explanations are tenable.

Meanwhile, we were somewhat surprised that social media use and addiction were not predictive of opinions on the fairness of using social media for job decisions. We expected that those who used social media more, and especially those who were addicted, would also be more likely to have inappropriate material on their social media profiles (e.g., Karl et al., 2010). Thus, we expected that these individuals would oppose the use of social media for job decisions. The fact that this was not supported suggests one of two things: (1) that those who use social media heavily vary in their opinion of whether the information they present online should be subject to public scrutiny, or (2) that among people who use social media heavily, there is a good amount of variance in their engagement in inappropriate behaviors, their ability to filter out their inappropriate behaviors, or both. This is a direction for future research.

Finally, one unexpected finding that merits mention is that the older participants in our sample were more likely to oppose the practice of using social media for job decisions. As all of the participants in our study were young adults, we examined age only as a potential covariate and did not make a priori predictions about with regard to its direction of influence. However, our correlational analyses give some explanation for this finding. More specifically, those who were older were more likely to endorse the hookup culture and were also more likely to be open to experience. Therefore, they may have had more experiences that, if posted online, could be potentially damaging to their job search. They also could have, because of their openness past experiences, more flexibility with regard what is acceptable to post online within the prevailing culture and a greater understanding that what is posted might be potentially damaging in the eyes of future employers. As these older adults are likely to be closer to entering the job market (or active in the job market), these social media/work issues might also be more salient and/or objectionable to them.

4.1. Limitations

Our study does have limitations that need mention. First, this study was conducted in a university setting; therefore, it is not known whether these results are applicable to the more general population of young adults. However, even if we captured only the sentiments of those who are in college preparing for their future careers, this is a large population of future employees that will, in the future, be weighing in on this issue. Additionally, we used a single-item measure to assess participants' opinions regarding the use of social media for job decisions. We acknowledge that a lengthier, multi-faceted measure might be more desirable, and a multi-item measure is another direction for future research. That said, we did find significant predictive relationships even with this single-item measure. Finally, we examined only a limited number of psychological characteristics that could be related to people's opinions of whether social media should be used for hiring and firing decisions, and these variables accounted for only a small amount of variance in this opinion. We acknowledge that there are other variables not explored in this study that could influence opinions on this issue, and we look to future research to explore this topic further.

4.2. Conclusion

As legal cases are hitting the courts, both private and public companies are beginning to recognize the importance of developing policies regarding the use of social media for hiring and firing decisions. Until now, the public perception of this practice has gone unexplored; however, as laws are supposed to be influenced by the voices of the constituents, it is important to examine what the voices of the upcoming generation of employees are saying about this practice. In our study, most young adults were opposed to using social media in hiring and firing decisions, and this was more common among those who were more open to experience, had little self-control, and were more accepting of the hookup culture. Perhaps these individuals have more to hide in their social media profiles; however, it could also be that these individuals are more accepting and embracing of the current culture, in which the sharing of all types of information (both appropriate and inappropriate) via social media is commonplace. As this generation of employees enters the job market, it will be interesting to see how social media laws and policies adapt to prevailing cultural attitudes. However, and most importantly, this generation of upcoming workers must be informed that regardless of their opinions of the fairness of these policies, as it currently stands, their short-term social media use could have a long-term effect on their future careers.

References

Andreassen, C. S., Torsheim, T., Brunborg, G. S., & Pallesen, S. (2012). Development of a Facebook addiction scale. *Psychological Reports, 110*, 501–517. http://dx.doi.org/10.2466/02.09.18.PR0.110.2.501-517.

Aubrey, J., & Smith, S. E. (2013). Development and validation of the endorsement of the hookup culture index. *Journal of Sex Research, 50*, 435–448. http://dx.doi.org/10.1080/00224499.2011.637246.

Bonds-Raacke, J., & Raacke, J. (2010). MySpace and Facebook: Identifying dimensions of uses and gratifications for friend networking sites. *Individual Differences Research, 8*, 27–33.

Costa, P. T., & McCrae, R. R. (1992). Four ways five factors are basic. *Personality and Individual Differences, 13*, 653–665.

DeBono, A., Shmueli, D., & Muraven, M. (2011). Rude and inappropriate: The role of self-control in following social norms. *Personality and Social Psychology Bulletin, 37*, 136–146. http://dx.doi.org/10.1177/0146167210391478.

Digman, J. M. (1990). Personality structure: Emergence of the five factor model. *Annual Review of Psychology, 41*, 417–440.

DiMarzo, G. M. (2012). Why can't we be friends? Banning student teacher communication via social media and the freedom of speech. *American University Law Review, 62 Rev.*, 123.

Duggan, M., & Brenner, J. (2013). The demographics of social media users – 2012. *Pew Internet & American Life Project*, <http://pewinternet.org/~/media/Files/Reports/2013?PIP_SocialMediaUsers.pdf>.

Duggan, M., & Smith, A. (2014). Social media update 2013. *Pew Internet Research*. <http://pewinternet.org/Reports/2013/Social-Media-Update.aspx>.

Fulmer, E. H. (2010). Privacy expectations and protections for teachers in the internet age. *Duke Law & Technology Review* (014).

Gailliot, M. T., & Baumeister, R. F. (2007). Self-regulation and sexual restraint: Dispositionally and temporarily poor self-regulatory abilities contribute to failures at restraining sexual behavior. *Personality and Social Psychology Bulletin, 33*, 173–186. http://dx.doi.org/10.1177/0146167206293472.

Gangadharbatla, H. (2008). Facebook me: Collective self-esteem, need to belong, and Internet self-efficacy as predictors of the Igeneration's attitudes toward social networking sites. *Journal of Interactive Advertising, 8*, 5–15.

Hidy, K. M., & McDonald, M. S. E. (2013). Risky business: The legal implications of social media's increasing role in employment decisions. *Journal of Legal Studies in Business, 18*, 69–88.

Jobvite. (2013). Social recruiting survey results. <http://web.jobvite.com/Q313_SocialRecruitingSurvey_LandingPage.html>.

John, O. P., & Srivastava, S. (1999). The Big-Five trait taxonomy: History, measurement, and theoretical perspectives. In L. A. Pervin & O. P. John (Eds.). *Handbook of personality: Theory and research* (Vol. 2, pp. 102–138). New York: Guilford Press.

Karl, K., Peluchette, J. V., & Schlaegel, C. (2010). Who's posting Facebook faux pas? A cross-cultural examination of personality differences. *International Journal of Selection and Assessment, 18*, 174–186.

Land v. L'Anse Creuse Public School, Bd. Of Edu., 2010 Mich. App. LEXIS 999 (Mich. Ct. App. 2010).

McCrae, R. R. (1996). Social consequences of experiential openness. *Psychological Bulletin, 120*, 323–337. http://dx.doi.org/10.1037/0033-2909.120.3.323.

McFarland, S. (2012, March 12). If you want a job, you may have to turn over your Facebook password. *Business Insider*. <http://www.businessinsider.com/empoyers-ask-for-facebook-password-2012-3>.

McNee, E. (2013). Disrupting the Pickering balance: First Amendment protections for teachers in the digital age. *Minnesota Law Review, 97 Rev.*, 1818.

Miller, R. A. (2011). Teacher Facebook speech: Protected or not? *Brigham Young University Education and Law Journal, 2011 Rev.*, 637.

Mulvey, T. (2013). SHRM Survey Findings: Social Networking Websites and Recruiting/Selection. <http://www.shrm.org/research/surveyfindings/articles/pages/shrm-social-networking-websites-recruiting-job-candidates.aspx>.

Nadkarni, A., & Hofmann, S. G. (2012). Why do people use Facebook? *Personality and Individual Differences, 52*, 243–249.

Ostrow, A. (2009, August 10). Facebook fired: 8% of US companies have sacked social media miscreants. *Mashable*. <http://mashable.com/2009/08/10/social-media-misuse/>.

Payne v. Barrow County School District, Civil Case No. 09CV-3038-X, (Super. Ct. Ga. 2009).

Pickering v. Bd. Of Ed., 391 U.S. 563 (1968). Retrieved from LexisNexis Academic Database.

Protivity. (2013). *2013 Internal Audit Capabilities and Needs Survey Report*. <http://www.protiviti.com/en-US/Documents/Surveys/2013-IA-Capabilities-Needs-Survey-Protiviti.pdf>.

Raphan, E., & Kirby, S. (2014). Policing the social media water cooler: Recent NLRB decisions should make employers think twice before terminating an employee for comments posted on social media sites. *Journal of Business & Technology Law, 9 Rev*, 75.

Sheldon, K. M., Abad, N., & Hinsch, C. (2011). A two-process view of Facebook use and relatedness need-satisfaction: Disconnection drives use, and connection rewards it. *Journal of Personality and Social Psychology, 100*, 766–775.

Smith, A. (2011, November 14). Why Americans use social media. *Pew Research Center*. <http://www.pewinternet.org/Reports/2011/Why-Americans-Use-Social-Media.aspx>.

Stinson, J. (July 8, 2014). Resource: Password protected: States pass anti-snooping laws. *USA Today*. <http://www.usatoday.com/story/news/nation/2014/07/08/stateline-password-online-privacy-laws/12353181/>.

Stoughton, J., Thompson, L., & Meade, A. W. (2013). Big five personality traits reflected in job applicants' social media postings. *Cyberpsychology, Behavior, and Social Networking, 16*, 800–805. http://dx.doi.org/10.1089/cyber.2012.0163.

Tangney, J. P., Baumeister, R. F., & Boone, A. L. (2004). High self-control predicts good adjustment, less pathology, better grades, and interpersonal success. *Journal of Personality, 72*, 271–324.

Vohs, K. D., Ciarocco, N. J., & Baumeister, R. F. (2005). Self-regulation and self-presentation: Regulatory resource depletion impairs impression management and effortful self-presentation depletes regulatory resources. *Journal of Personality and Social Psychology, 88*, 632–657. http://dx.doi.org/10.1037/0022-3514.88.4.632.

Weidner, N., Wynne, K., & O'Brien, K. (2012). Individual differences in workplace related use of internet-based social networking sites. In Schmidt, G. B. & Landers, R. N. (Eds.), *The impact of social media on work*. Symposium presented at the 2012 meeting of the society for industrial and organizational psychology. San Diego, California.

Annual Review of Sociology

The Demand Side of Hiring: Employers in the Labor Market

David B. Bills,[1] Valentina Di Stasio,[2] and Klarita Gërxhani[3]

[1]Department of Educational Policy and Leadership Studies, University of Iowa, Iowa City, Iowa 52242; email: david-bills@uiowa.edu

[2]Nuffield College, University of Oxford, Oxford OX1 1NF, United Kingdom; email: valentina.distasio@nuffield.ox.ac.uk

[3]Department of Political and Social Sciences, European University Institute, San Domenico di Fiesole 50014, Italy; email: klarita.gerxhani@EUI.eu

Annu. Rev. Sociol. 2017. 43:291–310

First published as a Review in Advance on May 10, 2017

The *Annual Review of Sociology* is online at soc.annualreviews.org

https://doi.org/10.1146/annurev-soc-081715-074255

Copyright © 2017 by Annual Reviews. All rights reserved

Keywords

labor markets, stratification, human capital, social capital, cultural capital, employer behavior

Abstract

Sociological research on labor markets has focused most of its attention on the supply side of the labor market, that is, the characteristics of job seekers and job incumbents. Despite its pivotal and we believe primary role in labor market processes, the demand side, in particular the hiring decisions made by employers, has received less attention. The employment relationship, however, comprises both the demand and supply sides, as well as the matching processes that bring these together. We consider the sociology of the demand side by considering three sources of information (human, social, and cultural capital) that employers charged with making hiring decisions seek out, as well as the mechanisms associated with each source. We conceptualize employers as active agents whose hiring behavior is both constrained and enabled by larger social, organizational, and institutional contexts. We call for a program of research that will lead to a fuller empirical and theoretical understanding of employer hiring behavior and its place in the stratifying of labor markets.

INTRODUCTION

How employers collect, interpret, and act upon information about job candidates is crucial to understanding how people get jobs. These jobs are, in turn, crucial to access to labor market rewards—income, benefits, career opportunities, and job security—and the processes by which people get jobs are fundamental to social stratification. Despite the clear importance of the demand side of labor markets, however, sociologists' understanding of employer hiring behavior has been piecemeal relative to our understanding of the characteristics of job seekers and job incumbents.

In our view, the employer or demand side is not simply an additive to more prevalent studies of the worker or supply side. We see the demand side—the recruitment and selection decisions made by employers—as more fundamental as a determinant of inequality, opportunity, and organizational attainments than the supply side. Labor is a derived demand. The creation of an employment relationship begins with an employer's decision to hire and ends with an enactment of that decision. As Jackson has stated, labor market inequalities "are the macrolevel result of a whole range of microlevel decisions by employers and prospective employees, and ultimately employers get the final say about which characteristics of employees are rewarded in the labor market" (Jackson 2007, p. 370).

We proceed as follows. We first establish our conceptualization of the role of employers in labor markets. We then identify three broad sources of information employers rely on when they make hiring decisions, with a focus on the mechanisms that come into play for each of these sources. We start with human capital, focusing in particular but not solely on the centrality of educational credentials in hiring behavior. Second, we review the literature on social capital, especially studies that examine how employers use social networks in their recruitment and screening decisions. Finally, we consider cultural capital, focusing on the apparently nonmeritocratic and less tangible criteria used by employers when staffing positions. Throughout the review, we are attentive to how employer assessments of human, social, and cultural capital vary across occupational, organizational, and institutional contexts. We conclude with some consideration of broad social changes that are likely to transform the way employers hire.

CONCEPTUALIZING EMPLOYER HIRING BEHAVIOR

Studies of employer hiring behavior date back many decades. Both institutional economists (Malm 1955, Rees & Shultz 1971) and community sociologists (Lynd & Lynd 1959, Warner 1963) examined the ecology of hiring within specific geographic boundaries. These studies are rarely read now, and for decades stratification research has been dominated by the use of data drawn from broadly representative samples of job holders (Blau & Duncan 1967, Warren et al. 2002). The ability to estimate statistical associations between occupational placement and ascribed and achieved statuses revolutionized the field of status attainment, but attainment researchers failed to develop a comparable understanding of the hiring behavior of employers. Typically, when researchers thought at all about the demand side, they inferred the preferences and practices of employers from observing the characteristics of job incumbents (i.e., individuals who have already been sorted into jobs) (see Fernandez & Weinberg 1997 for a critical discussion of posthire studies). In analyses that compare outcomes between job incumbents, demand-side and supply-side processes are necessarily conflated.

The new structuralists of the 1980s promised to right this imbalance by drawing attention to the characteristics of the firms, industries, and regions in which workers sought their statuses (Baron & Bielby 1980). In the end, however, the new structuralism never broke sufficiently with status attainment research to permit a rich understanding of the demand side. As two early proponents of the new structuralism later, somewhat regretfully observed, "the new structuralism for the most part links stratification outcomes to the very same industrial characteristics and organizational dimensions (and invokes largely the same arguments) as are emphasized by economists' theories of wage determination" (Baron & Pfeffer 1994, p. 191). What was missing from supply-side accounts was any appreciation of employers as active, decision-making agents. In contrast, our conceptualization of employer hiring behavior insists upon employer agency. We refer to a variety of organizational actors with the ability to make, or at least influence, decisions to bring new members into an organization. This is in sharp contrast to a model of workers simply converting their resources into socioeconomic outcomes.

By conceiving of employers as agents, we gain a deeper understanding of the microfoundations of labor market processes. That is, agency offers the link by which we can understand how such macroprocesses as national systems of skill development and certification (Culpepper 2003, Thelen 2004) and such meso-level processes as organizational, bureaucratic, and accountability structures (Dobbin et al. 1993, Reskin & McBrier 2000, Kalev et al. 2006) are enacted at the microfoundational level. These macro and meso levels have been extensively reviewed elsewhere (Reskin et al. 1999, DiTomaso et al. 2007, Stainback et al. 2010). For our purposes, this work provides an important context for understanding the scope conditions within which microfoundations are expressed. Rivera (2011, p. 72, emphasis original) proposed that a focus on the demand side allows us to study "*the process* of evaluation itself," and specifically how employers receive and assess information about candidates in the recruitment and selection of new hires. Thus, studying employer hiring behavior allows us to analyze the cues and information on candidates that employers trust and consider relevant, the motives underlying employers' recruitment and screening actions, and how employers' actions vary across organizational and institutional contexts. Compared to approaches that relate individual characteristics to occupational rewards (status attainment research), our focus on employers' agency deals explicitly with the gatekeeping role of employers and the decisions they make about who can—and, equally importantly, who cannot—enter the organization. Our focus on the demand side is consistent with a recent research agenda on organizational stratification that stresses the role of workplaces as inequality-generating settings (Stainback et al. 2010, Tomaskovic-Devey & Avent-Holt 2017). We concur with this perspective (which relies mainly on employee data) that the sorting and matching of people to jobs take place at the workplace. Our contribution is that the role of organizational agents in charge of recruitment and selection decisions is key to bringing the firms back in. In other words, occupational rewards and organizational-level outcomes are conditional on employers' agency during the hiring process.

A richer understanding of employer hiring behavior is also crucial because of emerging changes in how organizations fill vacancies. Following such path-breaking work as Doeringer & Piore (1971), sociologists in the 1980s produced solid research on internal labor markets (ILMs). Such arrangements are now less common. Whereas employers at one time filled vacancies primarily by promoting workers on the ascent from their initial port of entry, hiring is increasingly a transaction between an employer and an external candidate. Relatively few job vacancies are now filled from within. A valuable review by Cappelli & Keller (2014, pp. 312–13) on the "decline of the traditional model" has noted that

> Perhaps the most fundamental change has been the expansion of external hiring. With ports of entry no longer restricted to lower-level jobs, employers now hire into almost all kinds of jobs at all levels of the organization.... Whereas large employers in the United States filled only about 10% of their vacancies from outside in the period from World War II to the 1980s, current estimates suggest that they now fill over 60% of vacancies from outside.

This shift from hierarchy to market (Williamson 1981) makes the understanding of contingent decision making on the part of employers even more important. The trend toward external hiring means that employers will increasingly be required to select among candidates whose performance they have not had the opportunity to observe directly. As a greater share of labor market transactions must rely on proxies for worker skill and competence, assessing how employers make hiring decisions amid this greater uncertainty becomes even more important to understanding the functioning of the labor market.

We necessarily place some boundaries around our review. We focus on research that uses data collected directly from employers about actual or sometimes simulated hiring decisions. Given our emphasis on employer agency, we pay little attention to studies that merely analyze the correlates of employer practices (e.g., research on how characteristics of the last hired employee depend on organizational size or industrial sector). Instead, we give priority to research designs that make employer hiring behavior explicit. We are well aware that, in addition to the three forms of capital reviewed here, gender, race, and social class often play important roles in employer hiring decisions. However, because it has been reviewed capably elsewhere (Pager & Shepherd 2008) and has developed somewhat independently from the research of most interest to us, we mention research on employer discrimination mostly in passing. Where relevant, we signal empirical studies that have found evidence of homophily; double standards; or differential returns to human, social, or cultural capital along gendered, classed, or racial lines. Finally, we draw primarily on the sociological literature but include work from economics and psychology when it intersects with our interests.

HUMAN CAPITAL, SIGNALING, AND CREDENTIALISM

Our discussion of human capital centers on the role of educational credentials during the hiring process. Few analysts, including the founders of human capital, signaling, or credentialist theory (Becker 1964, Arrow 1973, Stiglitz 1975, Collins 1979), would claim that schooling is the only information presented by job seekers and evaluated by employers. Still, educational credentials figure prominently in all theories of job assignment, even as they adduce different mechanisms by which schooling, skills, and hiring are linked. Human capital theory adopts a learning mechanism, in which schooling teaches students something useful and adds value to potential employees. In contrast, signaling theory, which is both an extension of and a challenge to human capital theory, holds that schooling merely sorts students based on characteristics that are already present before the start of formal education (a selection mechanism). In this view, schooling serves as a sorting machine, the rationale being that schooling is a signal of unobservable abilities (e.g., willingness to learn, perseverance, motivation) supposedly correlated with job performance. Credentialist theory maintains that employers use educational credentials as a means of social closure, often without regard to the content of what schooling either inculcates or signals.

Human capital, signaling, and credentialist theories all assume that employers have reasons, perhaps not always clearly perceived or articulated, for making the hiring decisions that they do. One cannot, however, adjudicate these theories by appraising the characteristics of job seekers and neglecting the behavior of employers. Jackson et al. (2005, p. 10) note that "education has an effect in this regard only in so far as it is taken into account by employers (or their agents) in the decisions they make about hiring, retaining, promoting, etc." Still, empirical tests of these theories have been attempted mainly with supply-side data.

Insight into the characteristics of candidates that employers find most valuable is especially important given the increasing expansion and diversification of higher education. The importance of such emerging qualitative differences as college quality, sector, institutional prestige, or field of specialization—collectively known as the "horizontal dimensions of college education" (Gerber & Cheung 2008)—may well be on the rise in the minds of employers, possibly replacing more traditional markers of ascription such as gender or race or, more pessimistically, creating new avenues for educational stratification based on social class (Lucas 2001, Torche 2011). As Gerber & Cheung (2008, p. 313) noted, "data from employers assessing how they evaluate different types of the same credential might help resolve whether human capital, social capital, or signaling processes are at work."

SCHOOLING AND THE ACQUISITION OF SKILLS: LEARNING, SELECTION, OR NEITHER?

Theories of job assignment often overemphasize how much weight employers actually give to educational credentials in their hiring decisions, as schooling is but one of the screens that employers use to sort candidates. A series of papers by Bills (1988a,b, 1990, 1992b) examined how employers perceive the link between schooling and the acquisition of skills across six organizations in the United States, focusing on a variety of occupational positions. Bills reported that employers generally recognized a link between schooling and skill requirements on the job, although educational credentials alone were rarely used as the sole indicator of skills. Employers did not explicitly deny the importance of screening based on educational credentials. Nevertheless, they often discounted information obtained from educational credentials when other, more direct indicators of productivity such as job history data were available, suggesting that "perhaps we do live in a credential society, but educational credentials are not the only currency" (Bills 1988b, p. 87).

Miller & Rosenbaum (1997), based on in-depth interviews with a sample of employers of high school graduates in the Chicago area, found that employers consciously chose not to use information received from high schools. In their view, human capital and signaling theories are correct about the information that employers would like to use, but say little about the information that employers are willing to trust. Employers trusted their own judgments more than they trusted educational credentials. They routinely turned to networks, as "they trust those to whom they are connected" (Miller & Rosenbaum 1997, p. 513; see also the section titled Social Capital and Informal Recruiting, below).

More recent studies corroborate this, showing that at the sub-baccalaureate level, employers had only a slight preference for postsecondary credentials relative to high school qualifications and were rather indifferent to the institution attended by associate's degree holders (Deming et al. 2016, Deterding & Pedulla 2016). In particular, Deterding & Pedulla (2016) reported that employers did not draw meaningful distinctions between credentials earned at community colleges, a more expensive private for-profit institution (the fastest-growing segment of US postsecondary education), or even fictional institutions. The authors argued that this result challenges human capital and signaling theories of hiring and is consistent with Meyer's (1977) legitimation theory (see also Baker 2014): In an ever-expanding education system, employers rely on the legitimacy of educational credentials per se rather than on the quality of credentials that job candidates hold. Nunley et al. (2016) also questioned the validity of educational credentials as meaningful indicators of skills, based on the finding that employers hiring for business-related jobs did not prefer graduates with business degrees over graduates with other college majors. In sum, this literature does not deny the importance of skills for employers but questions the conditions under which employers use educational credentials as a meaningful indicator of skills, especially when other types of information are readily available [for example, occupational licenses (Deming et al. 2016)].

The Importance of Context

The employer studies reviewed so far have been conducted in the United States, a country with a generalist education system (Kerckhoff 1995, Torche 2011). In many European countries, including Germany, Switzerland, and the Netherlands, the education system has a tighter coupling between qualifications and the labor market, which is likely to affect employers' willingness to rely on educational credentials when hiring. Humburg & Van der Velden (2015) found that employers in nine European countries searched mostly for signals of occupation-specific human capital, such as relevant work experience or college major. Di Stasio & Van de Werfhorst (2016) analyzed country differences within Europe in more detail and showed that only Dutch employers relied predominantly on college major to sort applicants. British employers, instead, were more inclined to associate education with future trainability and preferred to rely on grades for their hiring decisions. Thus, common explanations of the informational value of credentials for employers require more specification, both in terms of conceptualization and measurement and in terms of the supply and demand conditions under which they apply (Bills 1992b).

Besides national institutions, within-country variation is also likely. Several studies have analyzed the hiring decisions of elite employers, on the assumption that educational credentials should be more relevant when stakes are high (Jackson 2009, Rivera 2011, Gaddis 2015). In her qualitative study of American investment banks, top-tier law firms, and management consulting firms, Rivera (2011) concluded that what matters to elite employers is not the length or content of education but rather its prestige. Competition is so intense that employers restricted their search to the most prestigious campuses (i.e., top-four superelite institutions), using prestige rankings as a shortcut for applicants' intelligence and ability to learn but also as a signal of their social skills, presentation style, and leadership potential. As Rivera (2011, p. 79) evocatively puts it, "the credential that employers valued was not the education received at a top school but rather a letter of acceptance from one." Using a field experiment with matched pairs of resumes, Jackson (2009) and Gaddis (2015) confirmed the importance of institutional prestige in employers' evaluations. Focusing on race, Gaddis (2015) found that credentials from an elite university increased the chances of a call-back for both white and black candidates, but ultimately, blacks with elite degrees did only as well as whites from less selective institutions, a clear sign of double standards.

Research on the contextual variation of employer hiring is still at an early stage. We hypothesize that differences between national institutions would especially affect hiring practices for lower-level bureaucratic positions, which often require standardized degrees earned in the formal education system (Brown 1995). By contrast, recruitment in professional and managerial labor markets is more likely influenced by professional associations, shared notions of cultural fit, or elite networks, which may vary less across national education systems. An interesting avenue for future research, which could put recent findings on the role of elite credentials in the hiring process (Rivera 2011, 2012a,b) into comparative perspective, would be to analyze elite labor markets in countries that do not have an education system stratified by institutional prestige (e.g., Germany or the Netherlands). Cross-national variation may be equally limited in the low-skilled labor market, where employers are governed mostly by incentives (e.g., wage subsidies for the long-term unemployed) and less by institutions (Bonoli & Hinrichs 2012). The state of the business cycle may also play a role. Indeed, Van Noy & Jacobs (2012) found that employers in Seattle (with a generally robust economy) valued associate degrees more than did employers in the more distressed Detroit economy. Finally, selection criteria may vary even across firms that seek to fill the same type of job. Di Stasio (2014) reports a large variation in the qualifications that were mentioned by employers as minimum entry requirements for jobs in the information technology sector. Coverdill & Finlay (1998, p. 122) reported that headhunters went as far as saying that "it may not be possible to speak of anything like a single set of criteria that guide the buying and selling of labor even within a narrowly defined labor-market context."

WHAT EMPLOYERS TRY TO AVOID: THE ROLE OF NEGATIVE OR STIGMATIZING INFORMATION

The focus on educational credentials has directed attention to the reasons why employers want to hire someone, but employers often engage in a search for "negative screens that can hurt but rarely help a candidate" (Bills 1990, p. 30). Employers sometimes seek indicators that serve to stigmatize job seekers, such as past episodes of absenteeism, lateness to work, job hopping, and unexplained gaps in one's resume (e.g., lengthy spells of unemployment). At times of educational expansion and differentiation, people with a mismatched job history or episodes of skill underutilization may carry a stigma, sometimes as scarring as a year of unemployment (Pedulla 2016; see also Nunley et al. 2016).

Even though job history data are at face value objective evaluation criteria, recalls or interpretation of past events may be biased against members of disadvantaged groups. For example, Pager & Karafin (2009) showed that positive experiences with black employees did not change employers' beliefs about black men in general. Those employees who did not conform to the stereotypical image of blacks lacking motivation or work ethic were viewed as the exception rather than the norm, leading the authors to coin the term Bayesian bigots.

The use of stigmatizing screens may be particularly prevalent in the low-wage, low-skilled labor market in which employers need quick shortcuts to sift through a pile of applicants (e.g., Pager & Karafin 2009, Bonoli & Hinrichs 2012). However, in these same markets, rejecting applicants based on stigmatizing screens such as a long-term unemployment spell may not be a feasible strategy, as the number of applicants may decrease drastically (Bonoli & Hinrichs 2012). There is also some evidence that employers reject stigma-bearing seekers even when the applicants are otherwise highly qualified. For example, in their comparative study of European college graduates, Humburg & Van der Velden (2015) found that deficiencies in academic, creative, interpersonal, organizational, and commercial skills were penalized to a greater extent than proficiency in these same areas was rewarded.

In general, research on negative screens is scant, other than studies focusing on discrimination on the grounds of gender or race (which fall outside the scope of our review). We note that employers do not always reject candidates with skill deficiencies or with negative or stigmatizing screens. Rosenbaum & Binder (1997) showed that employers took actions, sometimes costly ones, to cope with skill deficiencies. These actions included adjusting jobs to match employees' skills, investing in supervising new hires and in long-term linkages with schools to prescreen potential recruits. Employers' actions actively create labor market structures that increase their confidence in the information they receive from schools and teachers. In a related study, Miller & Rosenbaum (1997) reported that, despite the ready availability of school-based indicators of human capital such as grades, coursework, or school transcripts, employers did not use this information during the hiring process. They equally mistrusted information obtained from employment agencies, tests, or former employers or teachers as being potentially self-serving. In general, educational credentials are surprisingly often not the linchpin of labor market success. We develop this point further in the next section.

SOCIAL CAPITAL AND INFORMAL RECRUITING

To understand how social capital operates in the labor market from a demand-side perspective, scholars have looked at informal recruitment and the added value that social networks can bring to organizations. Research distinguishes between formal and informal methods of recruiting. Formal methods include public and private employment agencies, newspaper and online advertisements, and recruitment from schools and colleges. Informal methods include information from relatives, friends, acquaintances, employees, and other employers (Rees 1966). Sociologists, often influenced by Granovetter (1995), have been attentive to informal methods of recruitment and have considered such factors as loyalty, work ethic, and social skills (see Granovetter & Tilly 1988, Moss & Tilly 1996, Tilly & Tilly 1998, Marsden 2001). Informal recruitment takes place in a web of social networks characterized to a greater or lesser extent by trust, information flows, and reciprocity. In short, recruiting informally requires that employers activate their social networks (i.e., social capital). Lin's (2001, p. 19) understanding of social capital as "an investment in social relations with expected returns" succinctly captures the image of employer recruitment that we want to convey.

THE UNDERLYING MECHANISMS OF SOCIAL CAPITAL IN HIRING

When employers recruit informally, they draw upon different combinations of the social contacts of their incumbent employees along with their own business and professional contacts (Marsden & Gorman 2001). Several sociologists, and more recently economists, conducting organizational field studies have examined employers' reliance on referrals from incumbent employees (e.g., Fernandez & Weinberg 1997, Fernandez et al. 2000, Petersen et al. 2000, Neckerman & Fernandez 2003, Fernandez & Galperin 2014, Burks et al. 2015). They report that referrals increase the chances of candidates being invited for a job interview and lead to higher job offer rates compared to nonreferred applicants. Various mechanisms may explain the referral effect (Fernandez et al. 2000, Castilla et al. 2013): (*a*) Employers activating the social contacts of their own employees tap into a broader pool of candidates that would not be reached otherwise; (*b*) referrals are a prescreening strategy for the employer, as employees have an incentive to only vouch for friends or acquaintances who will ensure their own reputation protection; (*c*) referrers can provide their employers with hard-to-observe information about candidates (e.g., regarding soft skills, work ethic, past performance) while providing candidates with realistic information about the job and the organization; (*d*) social contacts tend to be homophilous and may indirectly signal candidates' competence, trust, status, or identification with the referrers; and (*e*) the presence of social ties between referrers and referred benefits the organizations directly owing to social processes that occur posthire, such as assistance during training. More recently, Fernandez & Galperin (2014) suggested an additional reason why employers may prefer referrals from incumbent employees, namely because they owe the candidate an extra look out of courtesy to the referrer.

In addition to trying to establish why employers may benefit from recruiting informally, research has also questioned whether the effect of social capital in the labor market is causal or spurious (Mouw 2003, 2006). Fernandez & Galperin (2014) leveraged multiple applications submitted by the same applicant to the same employer and found that employers are more likely to hire applicants when they have been referred than when the same persons apply without a network tie.

Besides employee referrals, other researchers have demonstrated that the social capital accessed via employers' own business and professional contacts is important, too (Petersen et al. 2000, Behrenz 2001, Marsden 2001, Pinkston 2012, Gërxhani & Koster 2015). These contacts are far from simple business transactions and are typically intensely social. Gërxhani & Koster (2015), for instance, showed how contacts can be established and mobilized in gatherings of professional and branch organizations, conferences, and online networks, but also in informal settings, such as while playing golf or at a charity gathering. Studies using laboratory settings to establish the mechanisms underlying employer hiring behavior (Brandts et al. 2010, Schram et al. 2010, Gërxhani et al. 2013) have consistently found that employers regularly forgo formal recruitment channels and choose to recruit through their own social networks. They do so even when they have to establish costly and anonymous social networks with other employers when sharing information about a candidate's trustworthiness in the past.

The value of social capital is reflected in employer hiring behavior not only as hiring through networks but also as hiring for networks (Erickson 2001). Theoretical and empirical research indicates that employers consider a candidate's networks as social credentials that can be exploited once the candidate is hired (Lin 2001, Marsden 2001). Moreover, a candidate's social capital explains part of the human capital effect on outcomes such as better jobs or occupational achievement (Erickson 2001, Völker & Flap 2004). Finally, in a recent quasi-experimental study on the role of social capital in academic recruitment, Godechot (2016) reported that having a supporter on a recruitment committee increases the chances of one's career advancement in academia. The mechanism underlying this process is held to be influence. As shown by Zinovyeva & Bagues (2015), however, an influential connection may in some cases be willing and able to pull a few strings for less qualified or less productive candidates, thus overshadowing the potential informational advantages that strong ties can deliver. In their study of academic committees in Spain, weak ties were both better informed about candidates and less biased than strong ties.

Network recruitment is often associated with social closure, as it tends to bias recruitment toward those who share a tie with incumbents, thus excluding outsiders. For example, the wrong-network hypothesis posits that minorities are cut off from job opportunities because they belong to impoverished networks. Petersen et al. (2000) showed that the disadvantage of ethnic minorities at point of hire is due to their lack of access to contacts that can vouch for them within the organization. Moreover, owing to the homophilous nature of social ties—referred candidates are often "a carbon copy of the employee referring them" (Waldinger 1997, p. 371)—network recruitment is said to reproduce the demographic characteristics of the existing workforce.

Some scholars have challenged the view that employee referrals, and informal recruitment more generally, are inherently exclusionary. Simulation studies by Rubineau & Fernandez (2015) reported that organizations can increase diversity by encouraging underrepresented groups to participate in referral programs. To the extent that members of ethnic minorities do refer coethnics through homophilous referral chains, employee referrals may even sustain workplace diversity (e.g., Fernandez & Fernandez-Mateo 2006). Of course, homophily in the employee referral process may still increase segregation, as shown by Fernandez & Sosa (2005) for a US call center. Finally, formal recruitment is not, in and of itself, free from bias or necessarily more meritocratic than informal recruitment. In the study of Neckerman & Kirschenman (1991), employers recruited selectively to avoid applications from poor black neighborhoods. By advertising jobs in specific local newspapers, they could target their ethnic population of choice, de facto engaging in address discrimination.

The Importance of Context

The relationship between institutions and recruitment channels has received little attention, rarely going beyond the role of formal institutions (e.g., Flap & Boxman 2001, Marsden 2001). Even then, the effect of formal institutions is usually measured indirectly through organizational characteristics such as size or sector (see Marsden & Gorman 2001 for an overview). Two organizational characteristics are particularly important when employers decide which recruitment channels to use. First, choice of recruitment method varies between public and private organizations. Public organizations are more visible to the broader public and to the environment within which they operate. Because of this, they are more subject to legitimacy issues and civil service regulations than are private organizations (Tolbert & Zucker 1983, Dobbin et al. 1988). As a result, public organizations are less inclined to use informal recruitment channels that typically do not offer equal opportunities to everyone (i.e., informal channels rely on social connections whose accessibility is not equally distributed) (Gërxhani & Koster 2015). Furthermore, larger organizations have more financial resources to invest in more costly recruitment methods (Boxman et al. 1994), as well as a higher likelihood of having a governance structure that aims at bureaucratic control and formalization (Marsden 1994). Small organizations, in contrast, are more likely to use informal, word-of-mouth methods of hiring the right person (Cassell et al. 2002, Gërxhani & Koster 2015).

The few employer studies that have looked at occupational variation in recruitment (Cohen & Pfeffer 1986; Boxman et al. 1994; Marsden 1994, 2001; Flap & Boxman 2001; Gërxhani & Koster 2015) have found that informal recruitment channels are used most often for risky occupations (e.g., with damage potential or training costs) and for managerial jobs. The rationale for employers to do so is that social networks provide them with more accurate information about prospective employees, thus minimizing the costs of hiring errors (see Jacobs 1981 for an important early statement on risky occupations).

Economists, in particular, have found evidence that the labor market situation at the time of recruitment is also important (Russo et al. 2000, 2001). Because in times of excess demand of labor it is more difficult to find candidates, employers activate different forms of social capital by investing in one or multiple recruitment methods, including informal ones. This is less likely in times of excess supply.

Overall, employers' reliance on formal or informal recruitment channels is not based purely on costs and benefits but is also affected by organizational and environmental factors (Marsden, 1994, 2001; Rynes & Cable 2003). For example, Di Stasio & Gërxhani (2015) found that referrals from employers' business contacts are important if formal educational qualifications are perceived by employers as noisy signals. This finding supports the argument that informal recruitment can represent a compensatory strategy that allows employers to rectify poor signaling (Marsden 2001) and illustrates the importance of the institutional context in which recruitment takes place (see also Rosenbaum et al. 1990, Miller & Rosenbaum 1997, Brinton & Kariya 1998). A full specification of these contextual conditions is still needed. In particular, more comparative research across different institutional and organizational contexts, where the focus is not only on formal institutions but also on informal institutions (i.e., social norms), would help clarify how employers contribute to the job match.

Moreover, there is a pressing need for more research on the contextual factors that may limit or enhance employers' discretion in personnel decisions, what Reskin (2000) called "the proximate causes of discrimination." These factors are not limited to recruitment channels but include anonymous application procedures (e.g., Goldin & Rouse 2000, Åslund & Skans 2012), the formalization of personnel practices (Reskin & McBrier 2000; Kmec 2005, 2006), and the presence of accountability structures (Dobbin et al. 1993, Kalev et al. 2006; see Stainback et al. 2010 for an extended review). Employers likely adopt bundles of recruitment and selection practices, freely mixing formal and informal methods (Reskin & McBrier 2000). Kmec (2006) speculates that hiring agents who recruit and hire applicants based on formal practices may, at the same time, also rely on informal recruitment and staffing practices. Recent research also points to the presence of clusters of methods, which transcends the traditional distinction between informal and formal practices (de Larquier & Marchal 2016). Understanding the interplay between formalization practices and informal processes is thus a promising area for future research.

CULTURAL CAPITAL AND OTHER NONCOGNITIVE CRITERIA

Cultural capital—those "institutionalized, i.e., widely shared, high status cultural signals (attitudes, preferences, formal knowledge, behaviors, goods and credentials) used for social and cultural exclusion" (Lamont & Lareau 1988, p. 156)—has been one of sociology's premier concepts for the past generation. It has, however, made fewer inroads into the empirical analysis of employer hiring behavior than might be expected. Fifteen years ago, Bills's (2003) review of the extant research on the relationships between schooling and job assignment found little research that examined how employers assess cultural capital when hiring. Fortunately, some recent scholarship has begun to change that. Classical sociological work on class reproduction (Bourdieu & Passeron 1977) has long argued that children from privileged classes develop a variety of social and cultural resources that are valued in the education system as well as in the labor market. Compared to children from less advantaged backgrounds, they have a natural familiarity with the informal codes and cultural experiences and preferences of the dominant classes and can mobilize this cultural capital to gain access to high-status jobs and organizational rewards. Only recently, though, have empirical studies analyzed how cultural capital is translated into occupational gains and particularly how gatekeepers (i.e., employers) rely on cultural capital for their hiring decisions.

CULTURAL CAPITAL AS MATCHMAKER: HIGHBROW CULTURE OR OMNIVOROUS TASTE?

Various qualitative studies have pointed to the importance of cultural similarity between applicants and evaluators, although the exact type of cultural resources being valued—whether an interest in arts, literature, sports, or some other leisure pursuit—varies across jobs and occupations (Erickson 1996; Turco 2010; Rivera 2011, 2012a,b). For example, in the leveraged buyout industry, recruiters look for confident and competent alpha males and consider knowledge of sports a highly valued cultural resource that facilitates bonding among predominantly male, white colleagues. Familiarity with football can be crucial to securing elite assignments. Although African American men can capitalize on their knowledge of sports to bond with majority colleagues, women are often disadvantaged by having less access to these cultural resources (Turco 2010).

In a qualitative study of hiring in elite professional services, Rivera (2015) describes hiring as a process of cultural matching. Her interviews and ethnography reveal that employers are not simply seeking competent colleagues but seek "new hires with whom they could envision themselves developing intimate relationships on and off the job" (Rivera 2015, p. 1352). This search for exciting "playmates" is so emotionally charged that Rivera compares it with dating. Evaluators unanimously stress the importance of fit: New hires are expected to blend in socially and culturally with existing employees. Fit is not a substitute for skills but an additional filter used to differentiate between similarly qualified applicants from elite, or even superelite (i.e., top-four), institutions. Managers use different criteria to assess fit, depending on the stage of the hiring process. During resume screening, evaluators measure cultural fit by seeking similarities between themselves and the candidates, which suggests that homophily, or better, "homocultural reproduction," also plays a role in the evaluation of merit (Rivera 2011, p. 88). In job interviews, fit is assessed by feelings of spark or chemistry that may arise during personal interactions. Extracurricular activities undertaken during college are an additional proxy for fit, particularly time- and resource-intensive activities that resonate with the "concerted cultivation" of white middle classes (Lareau 2003). These extracurricular activities are, de facto, a filter on socioeconomic status that reproduces race- and class-based stratification.

Jackson (2009, p. 687), based on a resume study, found that elite names change the meaning of other signals included in the application so that "two candidates might claim to have an interest in football, darts and socializing, but the social meaning of that statement will differ depending on whether the candidate is named Edward Acheson-Gray or Gary Roberts." This indicates another, more complex mechanism of class-based stratification implying different returns to leisure pursuits depending on social background. A more recent resume study conducted with top law firms in the United States revealed that men benefit more than women from displaying traditional signals of social class such as family names and lifestyle markers, possibly because higher-class women are perceived as less committed and unfit for such "intense, all or nothing occupations" (Rivera & Tilcsik 2016, p. 1122).

Although Bourdieu's work stresses the class dependency of highbrow cultural activities, Erickson (1996) notes that in most private sector organizations, the most widely useful cultural resource is cultural variety (i.e., the possession of a complex and varied cultural repertoire). According to Erickson (p. 219), "familiarity is the most portable and controllable form of culture, always available if needed or suppressible if inappropriate" and can be put to use strategically, for instance, to make a good impression during job interviews. Her study of the private contract security industry also shows that the type of cultural consumption that is considered valuable in such an organizational setting is not the highbrow genre stressed by Bourdieu but rather a specific type of business-related culture, displayed through sports talk, knowledge of good restaurants, and, especially, familiarity with magazines that simultaneously signal competence and businesslike orientation. Likewise, Koppman (2016), in her study of recruitment in advertising agencies, shows that, to enter creative occupations, the basis for cultural matching is omnivorous socialization: When hiring entry-level copywriters and designers, employers looked for signals of shared cultural omnivorousness, which was viewed as an indicator of creative potential and quest for knowledge. An important difference with the studies above is that employers were interested in a specific way of consuming culture (i.e., the possession of culturally omnivorous tastes, interests, and styles) rather than in any specific shared interest. In other words, in creative industries, "omnivorous taste itself is the shared interest" (Koppman 2016, p. 320).

The Importance of Context

In sum, these studies show that employers use cultural similarities as a marker of merit in settings where traditional markers of competence such as educational credentials have little relevance (e.g., advertising) or are so widespread that they become a noisy signal (e.g., elite professional services). As noted by Bills (2003, p. 451), though, "it seems unlikely, however, that employers can concentrate only on cultural capital (or its absence) when hiring for very specialized technical positions or that they are overly concerned about it when serving as gatekeepers for applicants of lower-status jobs." Indeed, employers in the low-wage, low-skilled labor market mentioned social and interpersonal skills among the most important qualities they look for in job applicants (Kirschenman & Neckerman 1991, Moss & Tilly 1996, Waldinger 1997). In particular, people skills such as the ability to interact with customers and coworkers as well as having the right attitude were high in demand by employers. However, Zamudio & Lichter (2008) interpret the fact that employers emphasize social skills even for entry-level, back-of-the-house jobs requiring little customer contact (e.g., housekeeper) as evidence that social skills are mentioned as a shorthand for selecting workers that can more easily be controlled. Thus, whereas elite employers rely on cultural capital as a signal of initiative, drive, and ambition to select future leaders, in the low-wage and low-skilled labor market, employers are concerned with avoiding conflict at the workplace and search for indicators of tractability, often using race or ethnic markers as a shortcut (Waldinger 1997, Zamudio & Lichter 2008).

More work that delves into how employers gather, interpret, and act upon the cultural capital and other intangible attributes presented by job candidates is badly needed, especially given the expansion of the service sector in which these qualities are most relevant (Jackson et al. 2005) and the increasing demand for social and personal skills even within occupational groups (Dörfler & Van de Werfhorst 2009). The lack of comparative studies of the role of cultural capital in hiring is particularly severe. We know of no research that examines national differences in how employers evaluate cultural capital when hiring.

THE CHANGING CONTEXT OF EMPLOYER HIRING BEHAVIOR

Hiring behavior is deeply embedded in social structure, much more than a simple bilateral exchange of human, social, or cultural capital between employers and applicants. We have portrayed employers here as agents with some degree of autonomy, but their agency is both constrained and enabled by an array of supply, demand, legal, organizational, institutional, interpersonal, and other social factors that collectively comprise the labor market. The broad ecology in which employer hiring behavior is embedded is changing quite rapidly, sometimes in directions of employers' own choosing but more often in ways beyond the control of even the most monopsonistic employers. These changes are affecting the form of the human, social, and cultural capital on which employers base their hiring decisions. We highlight here a few of the most significant. Each of these transformations opens up promising programs of research.

The Expansion of Education and the Proliferation and Diversification of Credentials

For many years, status attainment research was able to assume that schooling could be measured linearly. With the rapid expansion of higher education virtually everywhere, this assumption eventually became untenable. Researchers then shifted to categorical measures of schooling but still used categories that reflected participation in formal and traditional educational institutions, which were broadly construed as legitimate and chartered markers linking schools and workplaces. Recently, there has been a proliferation of credentials meant to certify skills acquired outside of formal schooling. Bestowed by diverse providers and variously known as certificates, badges, and microcredentials (M.R. Olneck, unpublished manuscript), these emerging signifiers of competence challenge the monopoly of credentials held by formal educational institutions and provide employers with a vast swath of hiring options and strategies. At this point, it is unclear if employers (as well as accreditors, licensure boards, and other labor market participants) will grant these microcredentials the legitimacy long accorded to formal educational credentials, which is why more research is needed in understanding whether and how employers value and select on these new forms of skill authentication.

Triadic Employment Relationships and Labor Market Intermediaries

As ILMs have grown less normative (Cappelli & Keller 2014), an increasing share of hiring is from the outside. The demise of ILMs does not mean that employers now encounter job candidates in any simple unmediated spot market. Rather, the new ecology of employer hiring behavior includes a vast assortment of labor market intermediaries (LMIs) intended to bring demand and supply together. As with many other organizational functions, recruiting, and even to some extent hiring itself, has become outsourced. The growing role of LMIs is producing what some have called triadic employment relationships (Khurana 2002, Bidwell & Fernandez-Mateo 2008, Bonet et al. 2013). Research on triadic employment relationships is still scarce. Future research should concentrate on the role of the different actors involved. For example, Fernandez-Mateo & King (2011) have shown that the placement consultants of a staffing firm made gender-biased choices in anticipation of a similar behavior by subsequent screeners (a mechanism they label anticipatory gender sorting).

Online Hiring: Beyond Paper People and Word of Mouth

Research on recruitment has traditionally distinguished between formal and informal methods, with information on potential hires either presented in a resume or conveyed through word of mouth. The diffusion of social media and online networking sites, such as LinkedIn, Twitter, and Facebook, raises questions about how information about vacancies is advertised as well as how information about candidates is received, actively sought, or validated. Although some studies have started to look into these issues (e.g., Manant et al. 2014, Baert 2015), more research is needed on this increasingly used channel of recruitment. Popular accounts have described the adoption of hiring algorithms in which individuals are matched to employers (or far more often, not matched) on the basis of the presence or absence of often mysterious keywords in their online resumes. Such algorithm-based hiring may bypass human decision making altogether. Thus, the advent of automated applicant-tracking systems (and the assortment of organizations rushing in to meet the various information needs of both employers and job seekers) potentially shifts the agency of the hiring decision from identifiable individuals to the hyperrationalized use of big data (Bidwell & Fernandez-Mateo 2008). This offers both possibilities to employers and constraints on their autonomy in ways that sociologists have barely begun to investigate. Moreover, as employers and employees increasingly interact on online platforms, the digital footprints of these interactions offer a new source of granular data to understand the interpersonal, relational nature of the hiring process (Ng & Leung 2015).

CONCLUSION: WHERE DO WE GO FROM HERE?

In an earlier review of employer hiring behavior, Bills (1992a) argued that we needed more empirical work on differences in screening across organizations, the interplay between screening and hiring criteria and labor market stringency, and how employers balance the various hiring criteria available to them. He added that we need an integrated theory that considers employers and job seekers simultaneously. Progress on these and other matters has been significant. We now know far more than we did a generation ago about how the hiring decisions of employers across many different contexts influence labor market outcomes. Of course, more work is still needed. We have suggested throughout this review promising avenues for future research.

Studying hiring behavior is difficult. Many employers are reluctant to provide access to curious social scientists. Some may have reasons to be evasive about their hiring behavior. Even well-meaning and accessible employers may be unreflective or uncertain of their own motivations and preferences. Still, we see many scientific benefits to be derived from an aggressive program of research on the strategic role of hiring in the labor market. Employer agency in hiring fills a gatekeeping role that can lead to closure and exclusion as easily as it can to merit and inclusiveness.

We need to learn how to collect and analyze data on the changing landscape on which employers now find themselves. Stratification researchers are not used to incorporating information on big data, social media, and LMIs into their research designs, but a full accounting of the stratifying role of employer hiring behavior is going to require this sort of ingenuity. The field would benefit from the collection of large-scale, representative survey data from employers to map their beliefs and actions. More effort might be expended on methodologies that bring us closer to the ground of employer behavior, such as case studies, (quasi-)experimental studies, the mining of personnel records or the text mining of job ads, and ethnography. Technological progress gives researchers ample opportunities to be creative in their data gathering efforts. For instance, the availability of crawling software makes it possible to collect data on job ads from multiple countries simultaneously (e.g., Kureková et al. 2015), and online labor markets such as Amazon Mechanical Turk are becoming increasingly acceptable platforms for running laboratory experiments that reach beyond undergraduate students (e.g., Rand 2012).

A hiring transaction is less a discrete event than it is one point on a continuum of events that bring demand and supply together. We need research that illuminates processes throughout this continuum. Attention should be given to data and research designs that can isolate demand-side mechanisms (Fernandez-Mateo & Fernandez 2016) from supply-side ones (Fernandez & Friedrich 2011, Ng & Leung 2015) and reveal whether stratified outcomes are due to the actions of screeners or the self-sorting of applicants (Fernandez & Campero 2016). An important avenue for future research would be to document how the joint actions of employers and job seekers construct applicant pools. We also need a richer understanding of employer behavior that takes place after the point of hire, a series of practices that Cappelli & Keller (2014) have characterized as talent management. Thus, we close with a call for theoretical and methodological innovation in research on employer hiring behavior and hope that our review will contribute to this innovation.

DISCLOSURE STATEMENT

The authors are not aware of any affiliations, memberships, funding, or financial holdings that might be perceived as affecting the objectivity of this review.

LITERATURE CITED

Arrow KJ. 1973. Higher education as a filter. *J. Public Econ.* 2:193–216

Åslund O, Skans ON. 2012. Do anonymous job application procedures level the playing field? *Ind. Labor Rel. Rev.* 65(1):82–107

Baert S. 2015. *Do they find you on Facebook? Facebook profile picture and hiring chances*. IZA Discuss. Pap. No. 9584, Inst. Study Labor, Bonn, Ger.

Baker D. 2014. *The Schooled Society: The Educational Transformation of Global Culture*. Stanford, CA: Stanf. Univ. Press

Baron JN, Bielby WT. 1980. Bringing the firms back in: stratification, segmentation, and the organization of work. *Am. Sociol. Rev.* 45:737–65

Baron JN, Pfeffer J. 1994. The social psychology of organizations and inequality. *Soc. Psychol. Q.* 57(3):190–209

Becker GS. 1964. *Human Capital*. New York: Natl. Bur. Econ. Res.

Behrenz L. 2001. Who gets the job and why? An explorative study of employers' recruitment behavior. *J. Appl. Econ.* 4:255–78

Bidwell M, Fernandez-Mateo I. 2008. Three's a crowd? Understanding triadic employment relationships. In *Employment Relationships: New Models of White-Collar Work*, ed. P Cappelli, pp. 142–78. Cambridge, UK: Cambridge Univ. Press

Bills DB. 1988a. Credentials and capacities: employers' perceptions of the acquisition of skills. *Sociol. Q.* 29(3):439–49

Bills DB. 1988b. Educational credentials and hiring decisions: what employers look for in new employees. *Res. Soc. Stratif. Mobil.* 7:71–97

Bills DB. 1990. Employers' use of job history data for making hiring decisions: a fuller specification of job assignment and status attainment. *Sociol. Q.* 31(1):23–35

Bills DB. 1992a. A survey of employer surveys: what we know about labor markets from talking with bosses. *Res. Soc. Stratif. Mobil.* 11:3–31

Bills DB. 1992b. The mutability of educational credentials as hiring criteria: how employers evaluate atypically highly credentialed job candidates. *Work Occup.* 19(1):79–95

Bills DB. 2003. Credentials, signals, and screens: explaining the relationship between schooling and job assignment. *Rev. Educ. Res.* 73:441–69

Blau P, Duncan OD. 1967. *The American Occupational Structure*. New York: Wiley

Bonet R, Cappelli P, Hamori M. 2013. Labor market intermediaries and the new paradigm for human resources. *Acad. Manag. Ann.* 7(1):341–92

Bonoli G, Hinrichs K. 2012. Statistical discrimination and employers' recruitment: practices for low-skilled workers. *Eur. Soc.* 14(3):338–61

Bourdieu P, Passeron J-C. 1977. *Reproduction in Education, Society and Culture*. Beverly Hills, CA: Sage

Boxman E, Flap H, Weesie J. 1994. Personeelsselectie door werkgevers: het belang van diepte-informatie. *Mens Maatsch.* 69(3):303–21

Brandts J, Gërxhani K, Schram A, Ygosse-Battisti A. 2010. Size doesn't matter! Gift exchange in experimental labor markets. *J. Econ. Behav. Organ.* 76:544–48

Brinton MC, Kariya T. 1998. Institutional embeddedness in Japanese labor markets. In *The New Institutionalism in Sociology*, ed. MC Brinton, V Nee, pp. 181–207. New York: Russell Sage Found.

Brown DK. 1995. *Degrees of Control: A Sociology of Educational Expansion and Occupational Credentialism*. New York: Teach. Coll. Press

Burks SV, Cowgill B, Hoffman M, Housman M. 2015. The value of hiring through employee referrals. *Q. J. Econ.* 130(2):805–39

Cappelli P, Keller JR. 2014. Talent management: conceptual approaches and practical challenges. *Annu. Rev. Organ. Psychol. Organ. Behav.* 1:305–31

Cassell C, Nadin S, Gray M, Clegg C. 2002. Exploring human resource management practices in small and medium sized enterprises. *Pers. Rev.* 31(6):671–92

Castilla EJ, Lan GJ, Rissing BA. 2013. Social networks and employment: mechanisms (part 1). *Sociol. Compass* 7(12):999–1012

Cohen Y, Pfeffer J. 1986. Organizational hiring standards. *Adm. Sci. Q.* 31(1):1–24

Collins R. 1979. *The Credential Society: An Historical Sociology of Education and Stratification*. New York: Academic

Coverdill JE, Finlay W. 1998. Fit and skill in employee selection: insights from a study of headhunters. *Qual. Sociol.* 21(2):105–27

Culpepper P. 2003. *Creating Cooperation: How States Develop Human Capital in Europe*. Ithaca, NY: Cornell Univ. Press

de Larquier G, Marchal E. 2016. Does the formalization of practices enhance equal hiring opportunities? An analysis of a French nation-wide employer survey. *Socioecon. Rev.* 14(3):567–89

Deming DJ, Yuchtman N, Abulafi A, Katz LF, Goldin C. 2016. The value of postsecondary credentials in the labor market: an experimental study. *Am. Econ. Rev.* 106(3):778–806

Deterding NM, Pedulla DS. 2016. Educational authority in the "open door" marketplace: labor market consequences of for-profit, nonprofit, and fictional educational credentials. *Sociol. Educ.* 89(3):155–70

Di Stasio V. 2014. Education as a signal of trainability: results from a vignette study with Italian employers. *Eur. Sociol. Rev.* 30(6):796–809

Di Stasio V, Gërxhani K. 2015. Employers' social contacts and their hiring behavior in a factorial survey. *Soc. Sci. Res.* 51:93–107

Di Stasio V, Van de Werfhorst HG. 2016. Why does education matter to employers in different institutional contexts? A vignette study in England and the Netherlands. *Soc. Forces* 95(1):77–106

DiTomaso N, Post C, Parks-Yancy R. 2007. Workforce diversity and inequality: power, status, and numbers. *Annu. Rev. Sociol.* 33:473–501

Dobbin F, Edelman L, Meyer J, Scott WR, Swidler A. 1988. The expansion of due process in organizations. In *Institutional Patterns and Organizations: Culture and Environment*, ed. LG Zucker, pp. 71–100. Cambridge, MA: Ballinger

Dobbin F, Sutton JR, Meyer JW, Scott R. 1993. Equal opportunity law and the construction of internal labor markets. *Am. J. Sociol.* 99(2):396–427

Doeringer PB, Piore MJ. 1971. *Internal Labor Markets and Manpower Analysis*. Lexington, MA: D.C. Heath

Dörfler L, Van de Werfhorst HG. 2009. Employers' demand for qualifications and skills: increased merit selection in Austria, 1985–2005. *Eur. Soc.* 11(5):697–721

Erickson BH. 1996. Culture, class, and connections. *Am. J. Sociol.* 102(1):217–51

Erickson BH. 2001. Good networks and good jobs: the value of social capital to employers and employees. See Lin et al. 2001, pp. 127–58

Fernandez RM, Campero S. 2016. Gender sorting and the glass ceiling in high-tech firms. *ILR Rev.* 70(1):73–104

Fernandez RM, Castilla EJ, Moore P. 2000. Social capital at work: networks and employment at a phone center. *Am. J. Sociol.* 105(5):1288–356

Fernandez RM, Fernandez-Mateo I. 2006. Networks, race, and hiring. *Am. Sociol. Rev.* 71(1):42–71

Fernandez RM, Friedrich C. 2011. Gender sorting at the application interface. *Ind. Relat.* 50(4):591–609

Fernandez RM, Galperin RV. 2014. The causal status of social capital in labor markets. *Res. Sociol. Organ.* 40:445–62

Fernandez RM, Sosa ML. 2005. Gendering the job: networks and recruitment at a call center. *Am. J. Sociol.* 111:859–904

Fernandez RM, Weinberg N. 1997. Sifting and sorting: personal contacts and hiring in a retail bank. *Am. Sociol. Rev.* 62:883–902

Fernandez-Mateo I, Fernandez RM. 2016. Bending the pipeline? Executive search and gender inequality in hiring for top management jobs. *Manag. Sci.* 62(12):3636–55

Fernandez-Mateo I, King Z. 2011. Anticipatory sorting and gender segregation in temporary employment. *Manag. Sci.* 57(6):989–1008

Flap H, Boxman E. 2001. Getting started: the influence of social capital on the start of the occupational career. See Lin et al. 2001, pp. 159–81

Gaddis M. 2015. Discrimination in the credential society: an audit study of race and college selectivity in the labor market. *Soc. Forces* 93(4):1451–79

Gerber TP, Cheung SY. 2008. Horizontal stratification in postsecondary education: forms, explanations and implications. *Annu. Rev. Sociol.* 34:299–318

Gërxhani K, Brandts J, Schram A. 2013. The emergence of employer information networks in an experimental labor market. *Soc. Netw.* 35:541–60

Gërxhani K, Koster F. 2015. Making the right move: investigating employers' recruitment strategies. *Pers. Rev.* 44(5):781–800

Godechot O. 2016. Getting a job in finance: the strength of collaboration ties. *Eur. J. Sociol.* 55:25–56

Goldin C, Rouse C. 2000. Orchestrating impartiality: the impact of "blind" auditions on female musicians. *Am. Econ. Rev.* 90(4):715–41

Granovetter M. 1995. *Getting a Job: A Study of Contacts and Careers*. Chicago: Univ. Chicago Press. 2nd ed.

Granovetter MS, Tilly C. 1988. Inequality and labor processes. In *Handbook of Sociology*, ed. NJ Smelser, pp. 175–271. Thousand Oaks, CA: Sage

Humburg M, Van der Velden R. 2015. Skills and the graduate recruitment process: evidence from two discrete choice experiments. *Econ. Educ. Rev.* 49:24–41

Jackson M. 2007. How far merit selection? Social stratification and the labour market. *Br. J. Sociol.* 58(3):367–90

Jackson M. 2009. Disadvantaged through discrimination: the role of employers in social stratification. *Br. J. Sociol.* 60(4):669–92

Jackson M, Goldthorpe JH, Mills C. 2005. Education, employers and class mobility. *Res. Soc. Stratif. Mobil.* 23:3–33

Jacobs D. 1981. Toward a theory of mobility and behavior in organizations: an inquiry into the consequences of some relationships between individual performance and organizational success. *Am. J. Sociol.* 87:684–707

Kalev A, Dobbin F, Kelly E. 2006. Best practices or best guesses? Assessing the efficacy of corporate affirmative action and diversity policies. *Am. Sociol. Rev.* 71(4):589–617

Kerckhoff AC. 1995. Institutional arrangements and stratification processes in industrial societies. *Annu. Rev. Sociol.* 21:323–47

Khurana R. 2002. Market triads: a theoretical and empirical analysis of market intermediation. *J. Theory Soc. Behav.* 32(2):239–62

Kirschenman J, Neckerman KM. 1991. "We'd love to hire them, but...": the meaning of race for employers. In *The Urban Underclass*, ed. C Jencks, PE Peterson, pp. 203–32. Washington, DC: Brookings Inst.

Kmec JA. 2005. Setting occupational sex segregation in motion: demand-side explanations of sex traditional employment. *Work Occup.* 32(3):322–54

Kmec JA. 2006. White hiring agents' organizational practices and out-group hiring. *Soc. Sci. Res.* 35(3):668–701

Koppman S. 2016. Different like me: why cultural omnivores get creative jobs. *Adm. Sci. Q.* 61(2):291–331

Kureková LM, Beblavý M, Haita C, Thum AE. 2015. Employers' skill preferences across Europe: between cognitive and noncognitive skills. *J. Educ. Work* 29(6):662–87

Lamont M, Lareau A. 1988. Cultural capital: allusions, gaps and glissandos in recent theoretical developments. *Sociol. Theory* 6(2):153–68

Lareau A. 2003. *Unequal Childhoods: Race, Class, and Family Life*. Berkeley: Univ. Calif. Press

Lin N. 2001. Building a network theory of social capital. See Lin et al. 2001, pp. 3–29

Lin N, Cook K, Burt RS, eds. 2001. *Social Capital: Theory and Research*. New Brunswick, NJ: Transaction

Lucas SR. 2001. Effectively maintained inequality: education transitions, track mobility, and social background effects. *Am. J. Sociol.* 106(6):1642–90

Lynd RS, Lynd HM. 1959. *Middletown: A Study in Modern American Culture*. New York: Harcourt Brace Jovanovich

Malm FT. 1955. Hiring procedures and selection standards in the San Francisco Bay Area. *Ind. Labor Relat. Rev.* 8:231–52

Manant M, Pajak S, Soulie N. 2014. Online social networks and hiring: a field experiment on the French labor market. MPRA Pap. No. 60403, Munich Pers. RePEc Arch. **https://mpra.ub.uni-muenchen.de/60403/**

Marsden PV. 1994. Selection methods in US establishments. *Acta Sociol.* 37(3):287–301

Marsden PV. 2001. Interpersonal ties, social capital and employer staffing practices. See Lin et al. 2001, pp. 105–27

Marsden PV, Gorman EH. 2001. Social networks, job changes, and recruitment. In *Sourcebook of Labor Markets: Evolving Structures and Processes*, ed. I Berg, AL Kalleberg, pp. 467–502. New York: Kluwer Acad./Plenum Publ.

Meyer JW. 1977. The effects of education as an institution. *Am. J. Sociol.* 83:55–77

Miller SR, Rosenbaum JE. 1997. Hiring in a Hobbesian world: social infrastructure and employers' use of information. *Work Occup.* 24(4):498–523

Moss P, Tilly C. 1996. "Soft" skills and race: an investigation of Black men's employment problems. *Work Occup.* 23(3):256–76

Mouw T. 2003. Social capital and finding a job: Do contacts matter? *Am. Sociol. Rev.* 68(6):868–98

Mouw T. 2006. Estimating the causal effect of social capital: a review of recent research. *Annu. Rev. Sociol.* 32:79–102

Neckerman KM, Fernandez RM. 2003. Keeping a job: network hiring and turnover in a retail bank. *Res. Sociol. Organ.* 20:299–318

Neckerman KM, Kirschenman J. 1991. Hiring strategies, racial bias, and inner-city workers. *Soc. Probl.* 38(4):433–47

Ng W, Leung MD. 2015. *For love or money? Gender differences in how one approaches getting a job*. IRLE Work. Pap. #103–15, Inst. Res. Labor Employ., Univ. Calif., Berkeley. **http://irle.berkeley.edu/workingpapers/103-15.pdf**

Nunley JM, Pugh A, Romero N, Seals RA. 2016. College major, internship experience, and employment opportunities: estimates from a résumé audit. *Labour Econ.* 38:37–46

Pager D, Karafin D. 2009. Bayesian bigot? Statistical discrimination, stereotypes, and employer decision making. *Ann. Am. Acad. Polit. Soc. Sci.* 621(1):70–93

Pager D, Shepherd H. 2008. The sociology of discrimination: racial discrimination in employment, housing, credit, and consumer markets. *Annu. Rev. Sociol.* 34:181–209

Pedulla DS. 2016. Penalized or protected? Gender and the consequences of nonstandard and mismatched employment histories. *Am. Sociol. Rev.* 81(2):262–89

Petersen T, Saporta I, Seidel MDL. 2000. Offering a job: meritocracy and social networks. *Am. J. Sociol.* 106:763–816

Pinkston JC. 2012. How much do employers learn from referrals? *Ind. Relat.* 51(2):317–41

Rand DG. 2012. The promise of Mechanical Turk: how online labor markets can help theorists run behavioral experiments. *J. Theor. Biol.* 299:172–79

Rees A. 1966. Information networks in labor markets. *Am. Econ. Rev.* 56(1–2):559–66

Rees A, Shultz GP. 1971. *Workers and Wages in an Urban Labor Market.* Chicago: Univ. Chicago Press

Reskin BF. 2000. The proximate causes of employment discrimination. *Contemp. Sociol.* 29(2):319–28

Reskin BF, McBrier DB. 2000. Why not ascription? Organizations' employment of male and female managers. *Am. Sociol. Rev.* 65(2):210–33

Reskin BF, McBrier DB, Kmec JA. 1999. The determinants and consequences of workplace sex and race composition. *Annu. Rev. Sociol.* 25:335–61

Rivera LA. 2011. Ivies, extracurriculars, and exclusion: elite employers' use of educational credentials. *Res. Soc. Stratif. Mobil.* 29:71–90

Rivera LA. 2012a. Diversity within reach: recruitment versus hiring in elite firms. *Ann. Am. Acad. Polit. Soc. Sci.* 639(1):71–90

Rivera LA. 2012b. Hiring as cultural matching: the case of elite professional service firms. *Am. Sociol. Rev.* 77(6):999–1022

Rivera LA. 2015. Go with your gut: emotion and evaluation in job interviews. *Am. J. Sociol.* 120(5):1339–89

Rivera LA, Tilcsik A. 2016. Class advantage, commitment penalty: the gendered effect of social class signals in an elite labor market. *Am. Sociol. Rev.* 81(6):1097–131

Rosenbaum JE, Binder A. 1997. Do employers really need more educated youth? *Sociol. Educ.* 70(1):68–85

Rosenbaum JE, Kariya T, Settersten R, Maier T. 1990. Market and network theories of the transition from high school to work: their application to industrialized societies. *Annu. Rev. Sociol.* 16:263–99

Rubineau B, Fernandez RM. 2015. Tipping points: the gender segregating and desegregating effects of network recruitment. *Organ. Sci.* 26(6):1646–64

Russo G, Gorter C, Schettkat R. 2001. Searching, hiring and labour market conditions. *Labour Econ.* 8:553–71

Russo G, Rietveld P, Nijkamp P, Gorter C. 2000. Recruitment channel use and applicant arrival: an empirical analysis. *Empir. Econ.* 25(4):673–97

Rynes SL, Cable DM. 2003. Recruitment research in the twenty-first century. In *Handbook of Psychology: Industrial and Organizational Psychology*, ed. WC Borman, DR Ilgen, RJ Klimoski, pp. 55–76. Hoboken, NJ: John Wiley and Sons

Schram A, Brandts J, Gërxhani K. 2010. Information, bilateral negotiations. *Eur. Econ. Rev.* 54:1035–58

Stainback K, Tomaskovic-Devey D, Skaggs S. 2010. Organizational approaches to inequality: inertia, relative power, and environments. *Annu. Rev. Sociol.* 36:225–47

Stiglitz JE. 1975. The theory of 'screening,' education, and the distribution of income. *Am. Econ. Rev.* 65:283–300

Thelen K. 2004. *How Institutions Evolve: The Political Economy of Skills in Germany, Britain, the United States and Japan.* New York: Cambridge Univ. Press

Tilly C, Tilly C. 1998. *Work Under Capitalism.* Boulder, CO: Westview Press

Tolbert PS, Zucker LG. 1983. Institutional sources of change in the formal structure of organizations: the diffusion of civil service reform, 1880–1935. *Adm. Sci. Q.* 28:22–39

Tomaskovic-Devey D, Avent-Holt D. 2017. Organizations and stratification: processes, mechanisms, and institutional contexts. *Res. Soc. Stratif. Mobil.* 47:1–5

Torche F. 2011. Is a college degree still the great equalizer? Intergenerational mobility across levels of schooling in the United States. *Am. J. Sociol.* 117(3):763–807

Turco CJ. 2010. Cultural foundations of tokenism evidence from the leveraged buyout industry. *Am. Sociol. Rev.* 75(6):894–913

Van Noy, Jacobs MJ. 2012. *Employer perceptions of associate degrees in local labor markets: a case study of the employment of information technology technicians in Detroit and Seattle.* CCRC Work. Pap. No. 39, Community Coll. Res. Cent.

Völker B, Flap H. 2004. Social networks and performance at work: a study of the returns of social capital in doing one's job. In *Creation and Returns of Social Capital: A New Research Program*, ed. H Flap, B Völker, pp. 172–96. London: Routledge

Waldinger R. 1997. Black/immigrant competition re-assessed: new evidence from Los Angeles. *Sociol. Perspect.* 40(3):365–86

Warner WL. 1963. *Yankee City.* New Haven, CT: Yale Univ. Press

Warren JR, Hauser RM, Sheridan JT. 2002. Occupational stratification across the life course: evidence from the Wisconsin Longitudinal Survey. *Am. Sociol. Rev.* 67:432–55

Williamson OE. 1981. The economics of organization: the transaction cost approach. *Am. J. Sociol.* 87(3):548–77

Zamudio MM, Lichter MI. 2008. Bad attitudes and good soldiers: soft skills as a code for tractability in the hiring of immigrant Latina/os over native Blacks in the hotel industry. *Soc. Probl.* 55(4):573–89

Zinovyeva N, Bagues M. 2015. The role of connections in academic promotions. *Am. Econ. J. Appl. Econ.* 7(2):264–92